Speech Recognition Algorithms Using Weighted Finite-State Transducers

Synthesis Lectures on Speech and Audio Processing

Editor
B.H. Juang, *Georgia Tech*

Speech Recognition Algorithms Using Weighted Finite-State Transducers
Takaaki Hori and Atsushi Nakamura
2013

Articulatory Speech Synthesis from the Fluid Dynamics of the Vocal Apparatus
Stephen Levinson, Don Davis, Scot Slimon, and Jun Huang
2012

A Perspective on Single-Channel Frequency-Domain Speech Enhancement
Jacob Benesty and Yiteng Huang
2011

Speech Enhancement in the Karhunen-Loève Expansion Domain
Jacob Benesty, Jingdong Chen, and Yiteng Huang
2011

Sparse Adaptive Filters for Echo Cancellation
Constantin Paleologu, Jacob Benesty, and Silviu Ciochina
2010

Multi-Pitch Estimation
Mads Græsbøll Christensen and Andreas Jakobsson
2009

Discriminative Learning for Speech Recognition: Theory and Practice
Xiaodong He and Li Deng
2008

Latent Semantic Mapping: Principles & Applications
Jerome R. Bellegarda
2007

Dynamic Speech Models: Theory, Algorithms, and Applications
Li Deng
2006

Articulation and Intelligibility
Jont B. Allen
2005

Speech Recognition Algorithms Using Weighted Finite-State Transducers
Takaaki Hori and Atsushi Nakamura

ISBN: 78-3-031-01434-5   paperback
ISBN:978-3-031-02562-4   ebook

DOI 10.1007/978-3-031-02562-4

A Publication in the Springer series
SYNTHESIS LECTURES ON SPEECH AND AUDIO PROCESSING
Lecture #10
Series Editor: B.H. Juang, *Georgia Tech*
Series ISSN
Synthesis Lectures on Speech and Audio Processing
Print 1932-121X   Electronic 1932-1678

Speech Recognition Algorithms Using Weighted Finite-State Transducers

Takaaki Hori and Atsushi Nakamura
NTT Communication Science Laboratories, NTT Corporation

SYNTHESIS LECTURES ON SPEECH AND AUDIO PROCESSING #10

ABSTRACT

This book introduces the theory, algorithms, and implementation techniques for efficient decoding in speech recognition mainly focusing on the Weighted Finite-State Transducer (WFST) approach. The decoding process for speech recognition is viewed as a search problem whose goal is to find a sequence of words that best matches an input speech signal. Since this process becomes computationally more expensive as the system vocabulary size increases, research has long been devoted to reducing the computational cost. Recently, the WFST approach has become an important state-of-the-art speech recognition technology, because it offers improved decoding speed with fewer recognition errors compared with conventional methods. However, it is not easy to understand all the algorithms used in this framework, and they are still in a black box for many people. In this book, we review the WFST approach and aim to provide comprehensive interpretations of WFST operations and decoding algorithms to help anyone who wants to understand, develop, and study WFST-based speech recognizers. We also mention recent advances in this framework and its applications to spoken language processing.

KEYWORDS

speech recognition, automaton, weighted finite-state transducer, Viterbi algorithm, decoder, optimization

Contents

Preface

The authors first encountered speech recognition technology nearly 20 years ago. At that time, rather than being a useful tool, it was one of the new technologies only available in a laboratory. In the 1990s, the authors began to study speech recognition. In those days, many researchers were focusing on large vocabulary tasks in the field of speech recognition. In addition to acoustic and language modeling with the aim of improving recognition accuracy, efficient decoding algorithms were also extensively investigated to find the most likely hypothesis in an enormous number of possible word sequences covering a large vocabulary.

One of the authors, Takaaki Hori, has long been engaged in the research and development of decoding algorithms for speech recognition, and has written many program codes for experimental and commercial systems. When he was a doctoral student, he started to develop a decoder for research purposes. The computer he used for the study was a Hewlett Packard C200/A4318A workstation with a 64MB memory, which was a high-end machine of the day. However, on this computer it was very hard to achieve speech recognition even for a 5,000-word vocabulary task. He dedicated himself to increasing the speed within the limitations of the available memory and computation power by introducing a new strategy and efficient algorithms. Despite his extensive experience of the development of speech recognition decoders in his company, he was fascinated to learn of the Weighted Finite-State Transducer (WFST) approach, which is the main theme of this book. The new approach was theoretically elegant and actually very fast as regards speech recognition decoding although it needed a large memory. Many heuristic techniques had already been proposed for conventional decoding approaches, but the essential components of these techniques were included more effectively and naturally in the WFST approach. At that time, he became convinced that the optimal decoding strategy for modern speech recognition was the WFST approach.

However, it was not easy to master the operations needed for manipulating WFSTs. Of course, there was publicly available software for the WFST operations, but they easily exhausted computer memory and often yielded unexpected results. The authors learned the basic properties of WFST operations from related papers and repeated trial and error, and they finally became skilled in the use of this approach. However, the time and memory consumption required for building and decoding with a WFST in large vocabulary tasks was often too large for the computers of the day to handle. For this reason, the authors concentrated on on-the-fly approaches that dynamically built a WFST during decoding and performed the decoding efficiently with a small memory footprint while maintaining its speed. As a result of this research and development, they finally wrote their own WFST tools and decoder.

In writing this book, the authors have tried to provide as much detail as possible about WFST-based speech recognition. It might be insufficient to provide a complete understanding of the theory and versatility behind the WFST, but those topics have been well covered in work cited in this book. Instead, this book focuses on its use in speech recognition. The authors' aim was to summarize the fundamentals and important techniques that have been proposed for speech recognition. They hope that this work will prove useful for researchers and developers who want to understand the WFST approach.

As regards writing this book, the authors would like to thank Prof. Shigeru Katagiri of Doshisha University for giving us the chance to work on WFST-based speech recognition when he was a superior of the authors at NTT Communication Science Laboratories, and Dr. Daniel Willett (currently at Nuance Communications, Inc.) and Dr. Yasuhiro Minami for teaching us about the initial implementation of a WFST-based speech recognition system. The authors would also like to thank Dr. Chiori Hori (currently at the National Institute of Information and Communications Technology, Japan) who helped with the task of extremely large vocabulary speech recognition using WFSTs, and Dr. Shinji Watanabe (currently at Mitsubishi Electric Research Laboratories), Dr. Mike Schuster and Dr. Erik McDermott (currently at Google Research) for the fruitful discussions in collaboration at NTT. Furthermore, the authors are deeply grateful to Prof. Masaki Kohda, emeritus professor at Yamagata University, Prof. Sadaoki Furui, emeritus professor of Tokyo Institute of Technology, and Prof. Yoshinori Sagisaka at Waseda University for giving them many opportunities and the motivation to pursue speech recognition research. Finally, they would like to express their gratitude to Prof. Biing-Hwang Juang at Georgia Institute of Technology for giving them the great opportunity to write this book.

Takaaki Hori and Atsushi Nakamura
January 2013

CHAPTER 1

Introduction

1.1 SPEECH RECOGNITION AND COMPUTATION

Nowadays, we often encounter computers that can recognize speech. For example, they allow us to enter text by speaking instead of typing. They may also answer our questions instead of a human operator. Furthermore, some applications not only recognize speech but also translate it into another language. Although the speech recognition performance of these devices is still inferior to the corresponding human ability, the result would be better if we were to speak clearly into the microphone.

Speech recognition is certainly useful when inputting linguistic information into computers. However, its mechanism is not widely known. How do such computers recognize speech? We can imagine software running on a computer (or a server) that somehow analyzes a speech signal and infers the spoken content. Indeed, a piece of software called a *speech recognition engine* is running on the computer, although the algorithm is complex, in which many computer science techniques have been adopted to achieve highly accurate and fast speech recognition.

Technically speaking, speech recognition is a process for converting a speech signal into text that corresponds to the spoken words conveyed by the signal. A speech recognition engine usually has knowledge sources related to acoustic and linguistic speech behavior and a program for finding the most likely word sequence for an input signal by referring to these knowledge sources. The knowledge representation and the algorithm are based on speech recognition technology that has been developed through a long history of research on speech and language processing.

Speech recognition research came to prominence in the 1970s and has made remarkable progress. The target of the research has been changing from the performance of simple tasks, e.g., isolated digit or word recognition, to more complex tasks, e.g., continuous word or sentence speech recognition. Before around 1990, the target vocabulary size of speech recognition systems was about 1,000 words. The vocabulary was limited to a specific domain such as resource management [FBP88, Pal89] and air travel information services [HGD90, PFFG90]. Because the computational power and memory size of computers were very limited, it was not easy to address large-scale tasks. However, most of the basic ideas appeared at that time and formed the basis of today's speech recognition technology. The statistical framework of speech recognition based on a source-channel model was established in the 1980s, and has become a standard framework that comprises Hidden Markov Models (HMMs) and n-gram language models [Jel98]. In this framework, speech signals are assumed to be signals encoded from message text, which are observed through a noisy channel.

Thus, the speech recognition process is interpreted as the *decoding* of the message text from the observed signals.

Speech signals vary in both the time and spectral domains. When single words are spoken, the speaking rate changes utterance by utterance. The spectral feature also changes depending on the speaker and the recording environment. Therefore, even if the same word is reiterated by the same speaker, the speech waveform will not completely match the previous one. In addition, the waveform does not contain explicit information indicating boundaries between phones, and it expands or contracts nonlinearly in the time domain. Consequently, it is computationally expensive to compare such speech waveforms properly while considering their nonlinear time alignment even in a simple task such as isolated word recognition. In spite of this large computational cost, speech recognition systems were usually required to work in real time on a standard computer. Thus, reducing the amount of computation required has constituted a major research topic in the speech recognition field.

In the early 1970s, the Dynamic Programming (DP) matching algorithm was introduced to compare a reference speech signal and an input speech signal efficiently for speech recognition [SC70, SC71].[1] Although DP matching is an efficient algorithm that takes account of the nonlinear expansion and contraction of a speech signal using the Dynamic Programming technique [BD62], it still needs $O(|R| \times |S|)$ computation where $|R|$ and $|S|$ denote the lengths of the reference speech signal R and input speech signal S, respectively. For isolated word recognition with vocabulary V, the complexity of recognizing one spoken word is $O(|V| \times |\bar{R}| \times |S|)$, where $|V|$ indicates the vocabulary size and $|\bar{R}|$ is the average length of the reference speech signals for words in V.

The research target moved to Large-vocabulary Continuous-speech Recognition (LVCSR) in the 1990s. In those days, a statistical framework was already a mainstream of the technology, which was supported with large corpora. The Defense Advanced Research Projects Agency (DARPA or ARPA) in the United States boosted the LVCSR research by undertaking a series of large projects. Many benchmark tests were designed for several speech data targets as part of those projects. Large-scale corpora were also intensively collected to acquire statistical models. Major institutes and universities developed LVCSR systems using these common corpora, and competed with each other in terms of accuracy and speed. As a result of those projects, the vocabulary size increased from 1K to 65K. The target speech included read newspaper articles, Broadcast News (BN), and conversational speech over the telephone. Thus, the LVCSR projects aimed to develop general-purpose speech recognition systems by targeting a wide range of speech less restricted by vocabulary, speaker, and speaking style.

However, LVCSR needs a large amount of computation. There is a synergy involved in increasing vocabulary size and coping with continuous speech that increases the complexity of decoding. A continuous speech signal contains multiple words, and yet the number of words and their time boundaries are unknown. Even if the system knows the number of spoken words to be L,

---

[1] DP matching is also called Dynamic Time Warping (DTW) in the speech recognition field.

the number of hypotheses to be compared is $|V|^L$ (e.g., $|V|^L = 10^{15}$ for $|V| = 1000$ and $L = 5$). If L is unknown, i.e., in a general case, it results in $\sum_{1 \leq l \leq \hat{L}} |V|^l$, where $\hat{L}$ is an assumed upper bound of L. If we enumerate all the hypotheses and compare each of them with the input signal, the complexity becomes $O(|V|^{\hat{L}} \times |\bar{R}| \times |S|)$. Thus, dealing with a large vocabulary in continuous speech recognition potentially has a large impact on the amount of computation required in the decoding process.

In fact, we do not have to enumerate all the hypotheses in LVCSR. Instead, we can use one-pass DP matching [BBC82, Ney84], which is a basic but efficient approach to continuous speech recognition. This method is called one-pass Viterbi algorithm when probabilistic models such as HMMs and an n-gram language model are used. If the HMMs are typical left-to-right type, i.e., each state has only one self loop and one exiting transition, the computational complexity is $O(|V|^{N-1} \times |\bar{M}| \times |S|)$, where n is typically 2 or 3 and $|\bar{M}|$ is the number of HMM states per word. Thus, the total computation is much less than that required for the full enumeration.

Most current speech recognition decoders are based on one-pass Viterbi algorithm (for details see Chapter 2). The algorithm ensures that the best hypothesis for a speech signal is found with given acoustic and language models. However, it is expensive to search for the best sequence of words from among a large vocabulary of over 10 thousand. In DARPA projects, search strategies for efficient decoding, which do not necessarily ensure the best hypothesis, were intensively investigated together with highly accurate acoustic and language models.

The most practical approach to reducing the computation for LVCSR decoding is to abandon the verification of all possible hypotheses. Beam search is the most popular method for reducing the number of hypotheses verified during the decoding process, which was originally introduced in 1976 [Low76]. With the Viterbi algorithm, partial sentence hypotheses are extended synchronously with time from the beginning of the speech. With the beam search, relatively unpromising partial hypotheses are selected and pruned out at each time frame. Those pruned hypotheses are no longer extended. As a result, the amount of computation can be reduced significantly since only some of the hypotheses are evaluated until the end of the speech signal. However, there is a potential risk that the correct partial hypothesis that would become the best sentence hypothesis may be lost by pruning. To eliminate such pruning errors, the beam search has been improved with various methods such as look-ahead techniques.

On the other hand, the efficient representation of speech information was also investigated to reduce redundant search space. A tree-organized lexicon was successfully introduced to represent the LVCSR search space. It is a data structure that shares pronunciation prefixes of the words in the vocabulary as a prefix tree (or called *trie*). This structure is also effective in alleviating the upswing of hypotheses at word boundaries, which causes an increase in pruning errors. Without a tree-organized lexicon, there are $|V|$ possible branches from the end of each word when extending partial hypotheses. By using this tree structure, the number of branches decreases to at most the number of phones.

In those days, many types of search strategies were proposed for the LVCSR decoding problem. Researchers studied stack-based time-asynchronous approaches such as A^*-stack decoding [KHG+91, Pau91], envelope search with fast acoustic match [GBM95], and multi-stack decoding [Sch00], which were performed differently from the one-pass Viterbi algorithm. The aim of these approaches was both to reduce the amount of computation and to reduce memory usage, which was severely limited by the computers of the day. The various search strategies for LVCSR are well summarized in [Aub02].

Multi-pass search strategies were also intensively investigated. A multi-pass search usually employs rough models that need less computation in the first pass to generate a set of promising sentence hypotheses. Then it uses detailed models that need more computation in the second pass to find the best hypothesis from a small set of hypotheses. Forward-backward search [ASP91], tree-trellis search [SH91], lattice N-best search [SC90], and word graph algorithm [ONA97] are well known approaches. If these strategies are not being used for real-time applications, more passes are often performed together with acoustic and language model adaptation to further improve the recognition accuracy. Moreover, the efficient representation of multiple hypotheses and how to generate a better set of hypotheses in the first pass decoding were also investigated at the same time.

Thus, LVCSR gradually became a reality along with the progress made on the decoding technology and the computational power of hardware. Some commercial software for LVCSR had appeared by the end of the 1990s, which was capable of taking dictation consisting of a user's continuous speech with a personal computer. For example, IBM ViaVoice(R) and Dragon Naturally Speaking(R) are representative products. These products made speech recognition familiar to the general population.

In the 2000s, a new paradigm has entered mainstream speech recognition technology, namely the Weighted Finite-State Transducer (WFST) approach and this is the main theme of this book. The framework was proposed in 1994 by researchers at AT&T Laboratories [PRS94, PR96, MPR96]. After that, the technique was improved steadily and has been considered the most efficient and theoretically elegant approach to the LVCSR decoding problem [MPR02]. Recently most major research institutes and universities have introduced this approach and are undertaking further investigations. In this book, we describe WFST-based speech recognition and related algorithms while paying attention to their differences when compared with traditional approaches.

1.2   WHY WFST?

Why has the WFST approach become so popular in the speech recognition field? The reason could be the aggressive use of *automata theory*. In other words, the approach takes full advantage of the theory over the whole decoding scheme, while ordinary methods have used it only in a limited fashion. This has resulted in an efficient and elegant framework.

The automata theory is the study of abstract computing devices, or *machines* [AHU74, HMU06], and it has long since entered mainstream computer science. Currently most undergraduate students on computer science courses learn the fundamentals of the automata theory, because automata are used in many areas such as logic circuits, data compression, cryptography, compilers, and natural language processing.

The theory has often been used to define a set of symbol sequences as a *language*, where an automaton is a way of expressing a grammar in formal language theory. For example, it is known that a regular grammar can be represented as a finite automaton, and a context free grammar can be represented as a push-down automaton.

A WFST is a sort of finite automaton. Roughly speaking, a basic finite automaton has a finite set of states and transitions between states. In ordinary finite automata, each transition has an input label. In addition, the WFST has an output label and a weight at the transition. Actually, finite automata were already being used in speech recognition before the WFST approach appeared. However, their use was limited to grammar representation as a language model. Since a WFST can to some degree represent relations between input and output strings with a weight that can correspond to some cost or probability, it can also represent relations between different levels of sequences such as HMM states, phones, and words in a unified framework.

In WFST-based speech recognition, WFSTs are typically used to represent an acoustic model, a pronunciation lexicon, and a language model, where the acoustic model WFST transduces an acoustic state sequence into a phone sequence, the lexicon WFST transduces a phone sequence into a word sequence, and the language model WFST transduces a word sequence into a sentence. Then those WFSTs are integrated by composition and optimization operations to a single WFST that directly transduces an acoustic state sequence into a sentence.

The WFST approach substantially increases the speed of most LVCSR tasks compared with traditional LVCSR approaches. Where does the difference come from? Two reasons have been presented.

1. Static search space organization
   The integrated WFST can be viewed as a large search network, where speech recognition is considered a search problem where the goal is to find a path that best matches the speech input signal in the network. This framework itself is trivial in speech recognition technology. But with the WFST approach, the network is statically stored in the memory known as *full expansion*, while traditional approaches usually construct a search network on demand, i.e., *dynamic expansion*. Since dynamic expansion needs a certain overhead during decoding, a fully expanded network is better for faster decoding.

2. Optimization of search network
   The search network often contains some redundancy. For example, different words often have partially the same pronunciation and are separately compared with the input signal during decoding. Such redundancy increases the number of computation and pruning errors when using a beam search. In the WFST framework, the redundancy can be removed by employing optimization operations such as *weighted determinization* and *minimization*.

These two factors seem to be straightforward. However, an LVCSR search network is extremely large and full expansion is actually almost impossible due to the limitation of memory size. Therefore, most LVCSR systems in the 1990s did not adopt such a full-expansion approach. The WFST accomplishes full expansion by reducing the search space with a series of optimization operations defined on WFSTs. As mentioned above, the optimization is also effective in removing the redundancy and accelerating the search process.

Some techniques have already been used to reduce the redundancy even in non-WFST-based speech recognition. A tree-organized lexicon is one example technique that is still widely used in many LVCSR systems. In the WFST framework, such a data structure can be constructed through optimization. The determinization of a WFST representing a lexicon yields a similar structure to the prefix tree. However, the use of determinization is not limited to a lexicon. It can be applied to the entire search space, i.e., the integrated WFST, and therefore the redundancy can be removed more rigorously. Minimization is also available, which corresponds to sharing suffixes in the case of a lexicon.

Accordingly, the WFSTs yield highly-optimized speech recognition for fast decoding. With the WFST framework, LVCSR has become faster than before. Currently real-time speech recognition with an extremely large vocabulary of over 1M words has become possible on a standard personal computer [HHM04, HHMN07].

On the other hand, simplicity is an attractive feature of the WFST framework. In speech recognition, the construction of a search network is basically performed by embedding knowledge sources hierarchically. An acoustic state sequence is embedded in a phone node, a phone node sequence is embedded in a word node, a word node sequence is embedded in a language model, and they are hierarchically associated to organize the search space. However, since the actual data structure and algorithm are not explicitly defined, the implementation changes system by system. The decoding program also depends strongly on the data structure, which is specialized for the models used in the system. Therefore, the system tends to have low expandability.

In the WFST framework, the construction process is well-explained by the composition operation that combines different levels of string-to-string relations. WFSTs of an acoustic state sequence to a phone sequence, a phone sequence to a word sequence, and a word sequence to a sentence can be composed into a single WFST of an acoustic state sequence to a sentence, which corresponds to the entire search space.

The actual construction of the integrated WFST, including the optimization step, is achieved by a small number of operations. Thus, the construction process is explicitly defined in this framework. The decoding program simply finds a path that best matches the speech input in the WFST, where the WFST acts as an interface between the models and the decoding program.

In addition, this framework provides the system with some flexibility. The decoder becomes more general. In many cases, we do not have to modify the decoding program to expand the function of the system. We can focus solely on the WFST that realizes the function. For example, speech-input language processing such as speech translation and speech summarization are possible without any decoder modification.

As stated above, the WFST framework has many advantages, which are well-explained based on the theory of finite automata. This is why it has become popular in the speech recognition field.

1.3   PURPOSE OF THIS BOOK

The main focus of this book is on the decoding problem in speech recognition. Currently the WFST approach is known as the most efficient way to solve this problem.

Many papers have been written by Mehryar Mohri, Fernando Pereira, and Michael Riley who are the pioneers of this approach. They have clearly described its concept, techniques, and impact on speech recognition speed. However, since these are academic papers for people that already know a lot about the decoding problem in speech recognition, they do not relate the basics of LVCSR decoding or provide details of WFST operations. On the other hand, major textbooks on speech recognition do not include WFSTs because they were written before the WFST became popular.

Algorithms of WFST operations can be found in some studies. But these studies explain an individual operation as a general-purpose tool and focus on its theoretical aspect. In fact, few studies deal with all algorithms for the construction of component WFSTs and the WFST operations needed for speech recognition. Thus, if we develop a WFST-based speech recognition decoder from scratch, we need to seek out and comprehend many studies. The purpose of this book is to aggregate such distributed knowledge, and provide comprehensive interpretations of WFST operations and decoding algorithms to help anyone who wants to understand, develop, and study WFST-based speech recognizers. In addition, we also mention recent advances in this framework and its applications to spoken language processing.

1.4   BOOK ORGANIZATION

This book is organized as follows. Chapter 2 defines speech recognition as a search problem. First, we present the statistical framework of speech recognition including Hidden Markov Models (HMMs) and n-gram language models. Second, we describe the one-pass Viterbi algorithm as a basic decoding method. Third, we introduce some traditional approaches to efficient large-vocabulary continuous-speech recognition based on search space optimization and aggressive pruning techniques. Finally, we show basic methods for generating multiple sentence hypotheses such as a word n-best list and a word lattice.

Chapter 3 introduces the WFST framework and its operations. First, we define the WFST and describe its basic properties. Second, we present algorithms that combine and optimize WFSTs such as composition, determinization, minimization, epsilon removal, and weight pushing.

Chapter 4 provides an overview of WFST-based speech recognition. First, we introduce the approach and show how speech recognition models can be represented in WFST form. Second, we describe the procedure for constructing a fully composed WFST for speech recognition. Finally, we show a Viterbi algorithm when using the composed WFST.

Chapter 5 focuses on dynamic decoders based on on-the-fly operations. Some efficient approaches have recently been proposed to achieve high-speed and memory-efficient decoding. We describe their principles and the algorithms.

Chapter 6 summarizes this book and offers some perspectives, where we overview recent applications based on the WFST approach with certain extensions. We also introduce publicly available software that can perform WFST operations and decoding.

Brief Overview of Speech Recognition

In this chapter, we review the statistical framework of speech recognition, which forms the basis of state-of-the-art speech recognition technology. However, since the technology is wide ranging and progressing day by day, this chapter focuses specifically on a standard approach to large vocabulary continuous speech recognition (LVCSR), which covers continuous-density hidden Markov models, context-dependent phone modeling, n-gram language models, time-synchronous Viterbi beam search, and an extension of the Viterbi search to obtain multiple sentence hypotheses. Since there have already been many studies describing these methods including the underlying theories and implementation techniques, we simply present a brief overview and emphasize how the methods are used in a Viterbi search.

The standard approach described in this chapter underlies WFST-based speech recognition. Therefore, it is important to describe the framework before explaining the WFST framework. However, use of WFSTs is not limited to the above methods. The standard approach is one example that can be optimized by using the WFST framework for fast decoding. Readers who are familiar with the basic methods may want to proceed directly to the next chapter (Chapter 3).

2.1 STATISTICAL FRAMEWORK OF SPEECH RECOGNITION

The statistical approach to speech recognition is modeled as a noisy channel model in information theory [JBM75, BJM83, Jel98]. The top of Fig. 2.1 shows a model where each component is associated with that of the general noisy channel model in the bottom of the figure. The speech recognition process includes a speaker and a speech recognizer. A word sequence W is generated in the speaker's mind (message source). W is then sent through an acoustic channel (noisy channel) consisting of a speech producer and an acoustic processor, where a speech producer vocalizes W and generates S in an acoustic environment, and the acoustic processor undertakes signal processing (and phone recognition) for S to obtain acoustic (and phonetic) features O. The linguistic decoder receives the feature vector sequence O and infers a word sequence $\hat{W}$ that is close to the original word sequence W. In this model, the acoustic processor and the linguistic decoder are included in the speech recognizer.

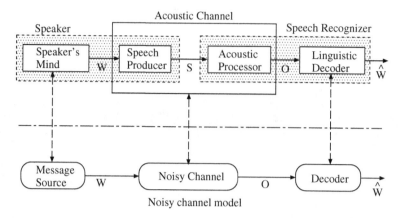

Figure 2.1: Speech recognition as a noisy channel model. The diagram shows speech recognition viewed as a noisy channel model where the upper block diagram is a model of speech recognition and the lower one is a general noisy channel model. The vertical bidirectional arrows indicate the correspondence between their components.

In accordance with information theory, the linguistic decoder finds the most likely word sequence $\hat{W}$ in a set of possible word sequences, $\mathcal{W}$, for a given input O, i.e.,

$$\hat{W} = \underset{W \in \mathcal{W}}{\operatorname{argmax}} P(W|O). \tag{2.1}$$

Eq. (2.1) is also known as the Bayes decision rule that classifies the input O into $\hat{W}$. This rule means that a class $\hat{W}$ with the maximum a posteriori (MAP) probability $P(\hat{W}|O)$ is selected and the probability of error minimized. Therefore, this type of decoder is often called a *MAP decoder*.

$P(W|O)$ can be rewritten by using the Bayes' theorem as

$$P(W|O) = \frac{p(O|W)P(W)}{p(O)}. \tag{2.2}$$

and therefore Eq. (2.1) can be rewritten as

$$\hat{W} = \underset{W \in \mathcal{W}}{\operatorname{argmax}} p(O|W)P(W), \tag{2.3}$$

where $p(O|W)$ is the acoustic likelihood of O for W, and $P(W)$ is the prior probability of W. Although $p(O)$ in Eq. (2.2) is the likelihood of O, $p(O)$ is not required in Eq. (2.3) since $p(O)$ is independent of $\hat{W}$. Accordingly, we consider only $p(O|W)P(W)$ when searching for $\hat{W}$, where $p(O|W)$ is calculated with an acoustic model and $P(W)$ is calculated with a language model.

Thanks to Bayes' theorem, we can separately model $p(O|W)$ and $P(W)$, and score possible sentence hypotheses for a given speech input using those models. The decoder works to find the

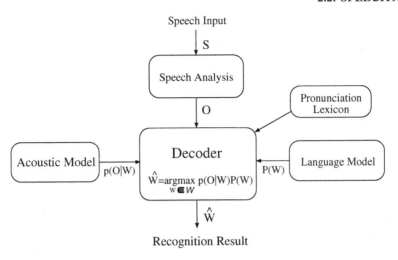

Figure 2.2: Continuous-speech recognition system

most likely sentence hypothesis $\hat{W}$ that is best-scored by the acoustic and language models, and output $\hat{W}$ as a recognition result.

According to this statistical approach, a standard speech recognizer is organized as in Fig. 2.2. Speech analysis is first applied to the speech waveform S and a feature vector sequence O is extracted. The decoder obtains the recognition result $\hat{W}$ using the acoustic model and the language model. Since the acoustic model is usually prepared phone by phone, the decoder requires a pronunciation dictionary that has a set of phone sequences for each word to calculate the acoustic likelihood of a word sequence. In the following sections, we review standard ways of constructing these components.

2.2   SPEECH ANALYSIS

The objective of speech analysis is to extract a feature vector sequence from an input waveform, where it is desirable that the vector space be separable into distinct phone categories.

Since speech is viewed as a piecewise stationary signal, speech analysis is generally performed at short time intervals of, e.g., 10 milliseconds, which is called a *time frame* or simply a *frame*. For example, after the A/D conversion of 16 kHz sampling, the signal is pre-emphasized with a high-pass filter and a power spectrum is derived using short-time Fourier analysis at each frame. Then, the 24 filter-bank amplitudes are obtained from the spectrum using mel filter-bank analysis, and their log amplitudes are transformed into Mel-Frequency Cepstral Coefficients (MFCCs) using the Discrete Cosine Transform (DCT). The first 12 MFCCs are used to form a feature vector together with the log power of the signal. A sequence of feature vectors is acquired continuously using a 25 millisecond time window and a 10 millisecond window shift. Moreover, their regression coefficients obtained from several consecutive feature vectors in the time domain can be appended to the original vector [Fur86]. The regression coefficients are called delta parameters and they capture the trend of each feature component in the time domain. The second or higher order regressions or delta of delta are also available. If we use 12 MFCCs, a log power, and their delta and delta-delta parameters, we obtain a 39-dimensional feature vector for each time frame.

Instead of appending the delta parameters, we can use a joint vector of consecutive feature vectors in the time domain, which is then transformed into a reduced dimensional space using Linear Discriminant Analysis (LDA) or Heteroscedastic LDA (HLDA) [KA98]. Moreover, posterior probabilities for phone categories are also used together with the above features, which can be derived using a Neural Network [HES00]. Actually there are many analysis methods and combinations for forming a feature vector, which aim to alleviate environment noise, channel distortion, and speaker variations, and improve class separability and recognition accuracy. Feature extraction methods are not detailed in this book. Readers interested in this topic should see [LJ93, HAH01].

The feature vector sequence O can be represented as $O = o_1, o_2, \ldots, o_t, \ldots, o_{T-1}, o_T$, where o_t is a feature vector at the t-th frame, and T stands for the length of O, i.e., the number of frames in O.

2.3   ACOUSTIC MODEL

In the statistical approach, an acoustic model is needed to calculate the acoustic likelihood $p(O|W)$. If O was obtained when a speaker uttered a word sequence W, the likelihood can be expected to take a large value. On the other hand, if the content uttered by the speaker differed from W, it can be expected to take a small value.

2.3.1   HIDDEN MARKOV MODEL

Hidden Markov Models (HMMs) are widely used as acoustic models for speech recognition. An HMM can be considered a model of a non-stationary information source. Speech is certainly a non-stationary signal and includes linguistic information in the time-variant spectral pattern. For example, if the speech includes the word "hello," the spectral pattern changes moderately according to the pronunciation /h/,/ae/, /l/ and /ou/. An HMM has a set of states, each of which is viewed as a statistically stationary signal source in speech recognition. The model generates a non-stationary signal by switching the state synchronously with the time frame.

Figure 2.3 shows a left-to-right HMM, which is often used to represent a phone model in speech recognition. The HMM has three states with a signal source and transitions to the state itself and the next state. The initial state of the HMM is the leftmost state from which it starts to output a feature vector sequence corresponding to the speech signal of the phone, and finally it arrives at the rightmost (final) state. At each time frame, one feature vector is output from a signal source associated with a state. A characteristic of HMMs is that it is assumed that the state transition process is probabilistic and therefore not observable, and only the output symbol (feature vector) sequence is observable. This property is effective as regards modeling speech signals, which vary depending on many factors including speakers, speaking styles, and recording environments.

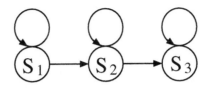

Figure 2.3: Left-to-right HMM: the nodes indicate the states with their identities S_1, S_2, and S_3, and the arcs represent possible state transitions.

Note that the HMM is a probabilistic model used for calculating acoustic likelihood, and therefore it actually does not output any symbols in a speech recognition system. The important thing is that when we observe a speech signal, we assume that the speech signal is generated by the HMM of W. Thus, $p(O|W)$ is considered the likelihood that the HMM of W generates the observed signal O.

An HMM is defined with the following parameters.

$$\theta = (\mathcal{S}, \mathcal{Y}, \mathcal{A}, \mathcal{B}, \Pi, \mathcal{F}), \tag{2.4}$$

where

$\mathcal{S}$: a set of states.

$\mathcal{Y}$: a set of output symbols.

$\mathcal{A}$: a set of state transition probabilities; $\mathcal{A} = \{a_{\sigma s}\}$; $a_{\sigma s}$ is the transition probability from state σ to state s, where $\sum_s a_{\sigma s} = 1$.

$\mathcal{B}$: a set of output probability distributions; $\mathcal{B} = \{b_s(x)\}$; $b_s(x)$ is the distribution of state s for $x \in \mathcal{Y}$, where $\int_{-\infty}^{\infty} b_s(x)dx = 1$.

Π: a set of initial state probabilities; $\Pi = \{\pi_s\}$; π_s is the initial probability at state s, where $\sum_s \pi_s = 1$.

$\mathcal{F}$: a set of final states.

These parameters can be estimated using training data based on the Maximum Likelihood criterion.

If we have N utterances O_1^N and the corresponding transcriptions W_1^N, a set of parameters for all phone HMMs, Θ, can be estimated so as to maximize the likelihood function,

$$L(\Theta|O_1^N, W_1^N) = \prod_{n=1}^{N} p(O_n|W_n, \Theta), \tag{2.5}$$

where $p(O_n|W_n, \Theta)$ denotes the acoustic likelihood as a function of Θ. We can also use other criteria such as maximum mutual information (MMI) and minimum phone error (MPE) [PW02]. Details regarding the training HMMs are not included in this book.

In speech recognition, the output symbols in Y correspond to feature vectors in a continuous space, and therefore an output probability distribution $b_s(x)$ is defined as a probability density function (PDF). This type of HMM is called a *Continuous Density* HMM to distinguish it from the standard type called a *Discrete* HMM.

2.3.2 COMPUTATION OF ACOUSTIC LIKELIHOOD

An acoustic likelihood with an HMM can be calculated by the Forward algorithm. Given an HMM $\mathcal{M}$ and a feature vector sequence $O = o_1, o_2, ..., o_T$, the acoustic likelihood $p(O|\mathcal{M})$, i.e., the likelihood that the model $\mathcal{M}$ generates O, is defined as

$$
\begin{aligned}
p(O|\mathcal{M}) &= \sum_S P(O, S|\mathcal{M}) \\
&= \sum_S \pi_{s_1} b_{s_1}(o_1) a_{s_1 s_2} b_{s_2}(o_2) \ldots a_{s_{T-1} s_T} b_{s_T}(o_T),
\end{aligned}
\tag{2.6}
$$

where S denotes a state sequence $s_1, s_2, \ldots, s_T$, and each state s_t means the state that arrives at time frame t.

In Eq. (2.6), $p(O|\mathcal{M})$ is obtained as the sum of the probabilities for all the state transition processes that can be followed in the HMM. The probability for a state sequence can be obtained by multiplying the state transition probabilities and the output probabilities according to the state sequence. However, summing up the probabilities by enumerating all possible state sequences needs an enormous amount of computation. To obtain the summation efficiently, we can utilize the Forward algorithm, which is calculated by the following recurrence formula.

$$
\begin{aligned}
\alpha(1, s) &= \pi_s b_s(o_1) \\
\alpha(t, s) &= \sum_{\sigma \in \mathcal{S}} \alpha(t - 1, \sigma) a_{\sigma s} b_s(o_t) \quad t = 2, \ldots, T
\end{aligned}
\tag{2.7}
$$

where $\alpha(t, s)$ means the forward probability that the model $\mathcal{M}$ outputs $o_1, \ldots, o_t$ and arrives at state s, i.e., $p(o_1, \ldots, o_t, s_t = s|\mathcal{M})$.

Finally, the acoustic likelihood is obtained at the final frame T

$$
p(O|\mathcal{M}) = \sum_{s \in \mathcal{F}} \alpha(T, s)
\tag{2.8}
$$

For the HMM in Fig. 2.3, the forward probability is calculated along a *trellis* as depicted in Fig. 2.4. The trellis represents a set of possible paths going from the initial state to the final state, each of which corresponds to a unique state sequence aligned to the feature vector sequence. On the trellis, the state transition probability at each arc and the output probability at each node are accumulated, and $\alpha(t, s)$ is obtained at each node by summing $\alpha(t - 1, \sigma) a_{\sigma s}$ from the preceding states $\sigma \in \mathcal{S}$, and then multiplying by $b_s(o_t)$.

When decoding for speech recognition, the Viterbi algorithm is usually used rather than the Forward algorithm. The Viterbi algorithm obtains the likelihood of the most likely state sequence

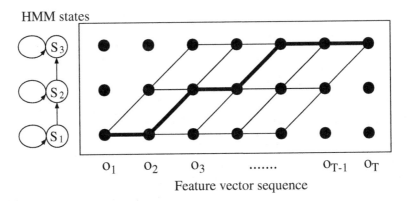

Figure 2.4: Trellis space in Forward algorithm and Viterbi algorithm

that outputs O as

$$
\begin{aligned}
\tilde{p}(O|\mathcal{M}) &= \max_S P(O, S|\mathcal{M}) \\
&= \max_S \pi_{s_1} b_{s_1}(o_1) a_{s_1 s_2} b_{s_2}(o_2) \ldots a_{s_{T-1} s_T} b_{s_T}(o_T),
\end{aligned}
\tag{2.9}
$$

This is similar to the likelihood given by the Forward algorithm, but $\sum$ in Eq. (2.6) is replaced with max.

As a result, (2.7) and (2.8) are also modified as

$$
\begin{aligned}
\tilde{\alpha}(1, s) &= \pi_s b_s(o_1) \\
\tilde{\alpha}(t, s) &= \max_{\sigma \in \mathcal{S}} \tilde{\alpha}(t - 1, \sigma) a_{\sigma s} b_s(o_t) \quad t = 2, \ldots, T
\end{aligned}
\tag{2.10}
$$

and

$$
\tilde{p}(O|\mathcal{M}) = \max_{s \in \mathcal{F}} \tilde{\alpha}(T, s),
\tag{2.11}
$$

where $\tilde{\alpha}(t, s)$ means the likelihood that the model $\mathcal{M}$ outputs $o_1, \ldots, o_t$ along the most likely state sequence arriving at state s. $\tilde{p}(O|\mathcal{M})$ is also called the *Viterbi score*.

In the example in Fig. 2.4, the Viterbi score $\tilde{p}(O|\mathcal{M})$ is the likelihood accumulated along the path indicated by the thick line, where we assume that the state sequence along the path maximizes $p(O, S|\mathcal{M})$. The path that gives the Viterbi score is called the *Viterbi path*.

In the algorithm, all the probabilities are usually accumulated in the log domain, i.e., the probabilities are transformed into their logarithmic values, and the multiplications are computed as additions in Eqs. 2.10 and 2.11. This is useful for avoiding the underflow or overflow caused by the iterative multiplications through the recurrence formula.

Although the Viterbi score is not the exact likelihood, there is no problem with respect to recognition accuracy if we use the Viterbi score instead of the Forward probability. Moreover, if we memorize the best preceding state obtained at each maximum selection in Eq. (2.10), the most likely state sequence that generates O can be obtained by back-tracking the best preceding states from the best final state selected in Eq. (2.11) at frame T. This procedure is basically the same as the Dynamic Programming [BD62], and can be directly applied to continuous speech recognition, which finds the most likely word sequence. The details are given in Section 2.7.1.

2.3.3   OUTPUT PROBABILITY DISTRIBUTION

HMMs can be classified depending on the type of output probability distribution at each state, i.e., Discrete and Continuous Density HMMs. The Discrete HMMs were originally incorporated into speech recognition, in which feature vectors are converted into discrete symbols using Vector Quantization (VQ). The output probability was defined on a finite set of code vectors. At present, Continuous Density HMMs are widely used, which have a multivariate Gaussian mixture density function at each state.

Given a feature vector x, a multivariate Gaussian mixture density at state i is computed as

$$b_i(x) \ = \ \sum_{m=1}^{M_i} c_{im} \mathcal{N}(x|\mu_{im}, \Sigma_{im}) \tag{2.12}$$

$$\mathcal{N}(x|\mu_{im}, \Sigma_{im}) \ = \ \frac{1}{\sqrt{(2\pi)^P |\Sigma_{im}|}} \exp\left\{ -\frac{1}{2}(x - \mu_{im})^t \Sigma_{im}^{-1}(x - \mu_{im}) \right\} \tag{2.13}$$

where $\mathcal{N}(x|\mu_{im}, \Sigma_{im})$ is the m-th Gaussian density. μ_{im} and Σ_{im} are the mean vector and the covariance matrix, respectively, and c_{im} is the branch probability (or *mixture weight*) for the m-th Gaussian, where

$$\sum_{m=1}^{M_i} c_{im} = 1.$$

M_i is the number of mixture components for state i, and P is the number of feature vector dimensions.

To reduce the required computation and the number of parameters, diagonal covariance matrices are often used, in which off-diagonal elements of the matrices are set at zero. In this case, each Gaussian distribution becomes dimensionally independent and can be computed as

$$\mathcal{N}(x|\mu_{im}, \Sigma_{im}) = \prod_{p=1}^{P} \frac{1}{\sqrt{2\pi \sigma_{imp}^2}} \exp\left\{ -\frac{(x_p - \mu_{imp})^2}{2\sigma_{imp}^2} \right\}, \tag{2.14}$$

where μ_{imp} and σ_{imp}^2 are the mean and variance values of the p-th dimension of the m-th Gaussian distribution. σ_{imp}^2 corresponds to the p-th diagonal element of the covariance matrix Σ_{im}.

For further refinement of output probability distributions, there are many attempts including Neural Networks [RMB+94, MDH09, PBA+11, SLY11].

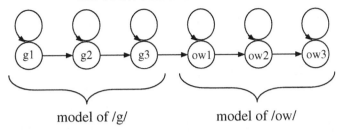

Figure 2.5: Example of an HMM for the word '*go*' synthesized with subword models of /g/ and /ow/

2.4   SUBWORD MODELS AND PRONUNCIATION LEXICON

Given an acoustic model of a word sequence W and a feature vector sequence O, the acoustic likelihood $p(O|W)$ can be computed using the Forward or Viterbi algorithm described in Section 2.3.2. However, since it is unrealistic to prepare a dedicated acoustic model for each possible word sequence beforehand, the acoustic models are usually made for subword units such as phones and syllables. An acoustic model of a word or a word sequence can be synthesized by concatenating the corresponding subword acoustic models. Using a pronunciation lexicon, we can synthesize an acoustic model for any word in the lexicon.

Let us assume that two subword models of phones /g/ and /ou/ are concatenated to synthesize a word model for '*go*' as in Fig. 2.5. The acoustic likelihood $p(O|W = \text{'}go\text{'})$ can be calculated using a Forward or Viterbi algorithm to obtain $p(O|\mathcal{M})$ with a single model $\mathcal{M}$. Figure 2.6 shows an example of a trellis for the concatenated HMMs, in which the thick line is assumed to be the Viterbi path. Since we know which state belongs to which phone unit, we can estimate the boundary time between units according to the Viterbi path. As shown with the dotted line in Fig. 2.6, the time frames 6 and 7 are estimated to be the most likely boundary between phones /g/ and /ou/. Thus, the Viterbi algorithm with concatenated HMMs is also useful for the automatic segmentation of speech.

2.5   CONTEXT-DEPENDENT PHONE MODELS

In acoustic modeling for speech recognition, finding a way to design a set of subword units has long been a problem. Currently, context-dependent phone units are widely used and known to be an effective approach. The acoustic property of each phone is not steady because it is influenced by adjacent or nearby phones. This property is called *coarticulation*. For example, the words *sketch* (/s k eh ch/) and *fox* (/f ao k s/) have the same phone /k/ but their sounds are not the same. Thus, a phone has allophones, and they are strongly dependent on the phone context. The incorporation of context-dependent phone HMMs is an effective solution for modeling such allophones accurately [Lee88].

A context-dependent phone is, for example, written as (s)k(eh) which stands for a phone /k/ preceded by /s/ and followed by /eh/. As in this example, a phone that depends on one preceding

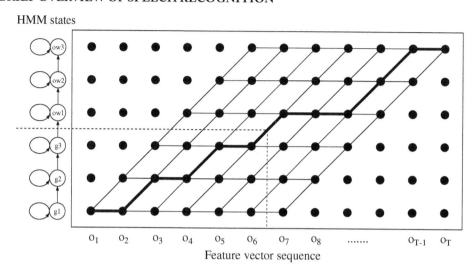

Figure 2.6: Trellis space with concatenated HMMs

phone and one succeeding phone is called a *triphone*. A word model for *'hello'* can be synthesized with triphone models of (sil)h(ae), (h)ae(l), (ae)l(ou), and (l)ou(sil), where we assume that the word is interposed in silences (/sil/). Actually, it is necessary to change the first and last triphones according to the preceding and succeeding words. Specifically, triphones exactly dealt with between words are called *cross-word triphones*. Although the use of cross-word triphones increases the complexity of decoding, it substantially reduces recognition errors compared with when using triphones only within each word model, i.e., using only a right-context-dependent model for the first phone and only a left-context-dependent model for the last phone.

In addition, it is known that *parameter tying* is effective when using context-dependent models. The number of context-dependent phones is much larger than the number of context-independent phones. If we have 40 phone units, we need to consider $40^3 (= 64,000)$ triphones. Although the triphones are not all usually observed in the training data, the number of observed triphones is much larger than the number of phones. In this case, it is difficult to reliably estimate all the parameters of the triphone models in the training phase because a triphone may have only a few samples with which to estimate the mean vectors, the covariance matrices, and the mixture weights for each HMM state. Furthermore, even triphones required in the decoding phase may not be seen in the training data. To recover such triphone models, there are certain techniques for interpolating the parameters by using those of other triphones that have similar coarticulations.

Tied-state triphone models have a shared output probability distribution at each state. A set of state clusters with a shared distribution is derived through decision tree construction [BdSG91, YOW94]. In [YOW94], a decision tree is constructed for each state of a phone model, where all the

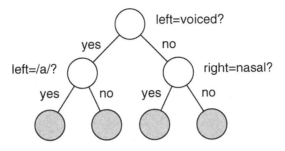

Figure 2.7: Phonetic decision tree

triphones derived from the phone model are assumed to have the same number of states as that of the phone model (usually three).

As a preliminary step in the decision tree construction, a set of phonetic questions is designed based on the phonetics of the language. Example phonetic questions are like "Is the left context a vowel?" or "Is the right context nasal?" A context class can be split by assigning each triphone in the class as either *yes* or *no* for a phonetic question as in Fig. 2.7.

First, only a root node is constructed, to which a context-independent phone class is assigned. At each step, a pair consisting of a leaf node and a question is selected from all possible pairs so that it gives the maximum gain of the expected log likelihood if the leaf node is split by the question. Then the best split is actually performed. The decision tree is grown by iterating this step. To stop the tree growth, we can utilize a condition where the number of frames assigned for each leaf node is greater than a predefined threshold. Since only a pair that satisfies this condition is adopted, the candidates for splitting gradually disappear, and finally the tree growth stops. We can also use a more advanced technique for this clustering step such as an MDL-based approach [SW00] and a variational Bayesian approach [WMNU04].

Thus, decision trees for all states of all phone models are constructed. A probability density function (PDF) such as a Gaussian mixture is assigned to each leaf node and the parameters are estimated with the training samples corresponding to the context class of the node. Using a decision tree, we can find appropriate PDFs for any triphone model (even for unseen ones) by tracing the tree from the root node to a corresponding leaf node. The transition probabilities are usually estimated independently of the context class.

2.6   LANGUAGE MODEL

A language model is used to define a set of word sequences that can be recognized by a speech recognizer. Some language models also give a probability or a weight to each word sequence as a language score. The score indicates the likelihood that the word sequence is uttered by an assumed

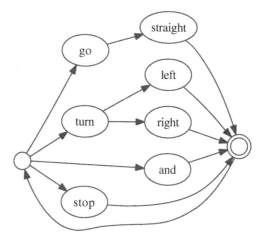

Figure 2.8: Finite-State Grammar

user. With the language model, ungrammatical and unlikely sentence hypotheses can be excluded from the speech recognition process, and therefore recognition errors are reduced.

Several types of language models are available for speech recognition, from which an appropriate model is selected according to the target domain.

2.6.1   FINITE-STATE GRAMMAR

If we can confidently predict what is spoken to the system, we can design the language model manually. In this case, Finite-State Grammars (FSGs) are usually used to design such hand-crafted models. There are several ways to represent FSGs. For example, a word is associated with a state, and a connection between words is represented as a state transition. Figure 2.8 shows an example of an FSG that represents a set of voice commands to a robot, where each node has a word label and directed arcs between nodes represent possible state transitions. A node without a word label indicates a null state. In the grammar, the leftmost node indicates the initial state and the double-circled node indicates the final state. This grammar accepts, for example, the utterance "*go straight and turn left.*" Accordingly, FSGs are useful for building speech recognizers for small tasks.

There is another style of FSGs in which the word label is assigned to each transition. It is known that an FSG with state labels is mutually convertible with an equivalent FSG with transition labels.

2.6.2   N-GRAM MODEL

The target domain of LVCSR is a naturally spoken language. Therefore, it is no longer realistic to build the language model by hand using highly specialized knowledge about the language. In most LVCSR tasks, spontaneous speech needs to be recognized that is often ungrammatical and unpredictable. Statistical language models are effective for recognizing such spontaneous speech, because they can be acquired from a large text corpus including many example sentences. In such statistical models, n-gram models are widely used for LVCSR.

An n-gram model is known as a simple and the most effective statistical language model for LVCSR. The term "n-gram" stands for a subsequence of n items in a sequence. The n-gram is referred to as a *unigram* when $n = 1$, a *bigram* when $n = 2$, and a *trigram* when $n = 3$. In the n-gram language model for speech recognition, the item corresponds to a word and n is usually 3 (trigram) or 4.

Let us denote a sequence of M words $w_1, w_2 \ldots, w_M$ as w_1^M and a subsequence from the i-th word to the j-th word as w_i^j. Given a word sequence $W = w_1^M$, the prior probability of W can be factored as

$$
\begin{aligned}
P(W) &= P(w_1)P(w_2|w_1)\ldots P(w_M|w_1^{M-1}) \\
&= \prod_{m=1}^{M} P(w_m|w_1^{m-1}).
\end{aligned}
\tag{2.15}
$$

In the n-gram model, the conditional probability $P(w_m|w_1^{m-1})$ is approximated with the n-gram probability $P(w_m|w_{m-n+1}^{m-1})$ by truncating the history of w_m into the previous $n - 1$ words. Note that w_{m-n+1} is assumed to be null if $m - n + 1 < 1$, e.g., $P(w_1|w_{-1}w_0) = P(w_1)$ when $n = 3$.

Accordingly, the prior probability of W is given as

$$
P(W) = \prod_{m=1}^{M} P(w_m|w_{m-n+1}^{m-1}),
\tag{2.16}
$$

where it is assumed that the occurrence probability of a word is dependent only on the previous $n - 1$ words.

It is also known that the n-gram model is equivalent to the $(n - 1)$-order Markov model in which a state of the Markov model corresponds to an $(n - 1)$-word history, and a state transition has an n-gram probability conditioned by the history of the source state.

Let w_1^n be an n-gram of words, $w_1, \ldots, w_n$. The maximum likelihood estimate of the n-gram probability can be obtained using a text corpus as

$$
P(w_n|w_1^{n-1}) = \frac{C(w_1^n)}{C(w_1^{n-1})},
\tag{2.17}
$$

where $C(w_1^m)$ is the number of occurrences of the n-gram, w_1^n, in the text corpus. However, since the corpus is finite even if we try to collect as much of it as we can, many n-grams may have zero or very few occurrences. Some important n-grams may happen to be unobserved in the corpus. Such n-grams are not well estimated statistically, and therefore not correctly recognized. Back-off smoothing described in the next section is a useful technique that can be used to mitigate such a sparse data problem for n-gram models.

2.6.3   BACK-OFF SMOOTHING

Back-off smoothing is a technique for estimating the probabilities of unseen n-grams using $(n-1)$-gram probabilities [Kat87]. The idea is basically to discount a part of probability of observed n-grams and redistribute the part to the unseen n-grams so that their probabilities are proportional to the corresponding $(n-1)$-gram probabilities.

The n-gram probability of w_1^n based on back-off smoothing has the following form:

$$P(w_n|w_1^{n-1}) = \begin{cases} P^*(w_n|w_1^{n-1}) & \text{if } C(w_1^n) > 0 \\ \alpha(w_1^{n-1})P(w_n|w_2^{n-1}) & \text{if } C(w_1^n) = 0 \end{cases} \tag{2.18}$$

where $P^*(w_n|w_1^{n-1})$ is the discounted n-gram probability, and $alpha(w_1^{n-1})$ is called a back-off coefficient when applying the $(n-1)$-gram probability for unseen n-grams, i.e., $C(w_1^n) = 0$.

Based on the discounted probabilities, the back-off coefficient is calculated as

$$\alpha(w_1^{n-1}) = \frac{1 - \displaystyle\sum_{w_n:C(w_1^n)>0} P^*(w_n|w_1^{n-1})}{1 - \displaystyle\sum_{w_n:C(w_1^n)>0} P^*(w_n|w_2^{n-1})}. \tag{2.19}$$

The numerator means the sum of unseen n-gram probabilities in the context of w_1^{n-1}. The denominator is the normalization factor.

There are several methods for estimating the discounted probabilities. With Good-Turing discounting [Kat87], $P^*(w_n|w_1^{n-1})$ is estimated as

$$P^*(w_n|w_1^{n-1}) = \frac{C^*(w_1^n)}{C(w_1^{n-1})} \tag{2.20}$$

using the modified count

$$C^*(x) = (C(x) + 1)\frac{N_{C(x)+1}}{N_{C(x)}}, \tag{2.21}$$

based on the Good-Turing estimate [Goo53], where N_r means the number of words that occurred in the corpus exactly r times. This estimate can be derived from Zipf's law.[1] The modified count is applied only when $C(x)$ is small, e.g., $C(x) \le 5$ in [Kat87].

2.7   DECODER

As mentioned in Section 2.1, the decoder is a computer program that searches for the most likely sentence hypothesis $\hat{W}$ based on Eq. (2.3) using an acoustic model and a language model. However,

---

[1]An empirical law observed in many types of data, in which the relative frequency of the k-th most frequent component is proportional to $1/k$.

it is difficult to compute Eq. (2.3) directly because a huge computation is needed for enumerating all possible sentence hypotheses W and calculating $p(O|W)P(W)$ for each W. It is well known that the one-pass Viterbi algorithm is effective for finding the best hypothesis $\hat{W}$.

2.7.1 VITERBI ALGORITHM FOR CONTINUOUS SPEECH RECOGNITION

Assuming that each sentence hypothesis W is a sequence of M_W words, $w_1, \ldots, w_{M_W}$, the likelihood of Eq. (2.3) can be factored into word-level scores as

$$
\begin{aligned}
\hat{W} &= \underset{W \in \mathcal{W}}{\mathrm{argmax}} \left\{ \sum_{S \in \mathcal{S}_W} p(O, S|W)P(W) \right\} \\
&= \underset{W \in \mathcal{W}}{\mathrm{argmax}} \left\{ \sum_{S \in \mathcal{S}_W} \prod_{m=1}^{M_W} p(o_{t_{m-1}+1}^{t_m}, s_{t_{m-1}+1}^{t_m}|w_m) P(w_m|w_1^{m-1}) \right\},
\end{aligned} \tag{2.22}
$$

where $p(o_t^\tau, s_t^\tau|w)$ denotes the likelihood that the model of word w generates the speech segment $o_t \ldots o_\tau$ along the state transition process $s_t \ldots s_\tau$. $\mathcal{S}_W$ denotes a set of possible state sequences for W. t_m represents the ending frame of word w_m, which is determined by the state sequence S, i.e., the requirement that s_{t_m} is a final state in the model of w_m and s_{t_m+1} is an initial state in the model of w_{m+1} is satisfied. Here we assume $t_0 = 0$.

In the case of Viterbi algorithm, the best word sequence $\hat{W}$ is obtained along the best state sequence as

$$
\begin{aligned}
\hat{W} &= \underset{W \in \mathcal{W}}{\mathrm{argmax}} \left\{ \max_{S \in \mathcal{S}_W} p(O, S|W)P(W) \right\} \\
&= \underset{W \in \mathcal{W}}{\mathrm{argmax}} \left\{ \max_{S \in \mathcal{S}_W} \prod_{m=1}^{M_W} p(o_{t_{m-1}+1}^{t_m}, s_{t_{m-1}+1}^{t_m}|w_m) P(w_m|w_1^{m-1}) \right\} \\
&= \underset{W \in \mathcal{W}}{\mathrm{argmax}} \left\{ \max_{T \in \mathcal{T}_W} \prod_{m=1}^{M_W} \tilde{p}(o_{t_{m-1}+1}^{t_m}|w_m) P(w_m|w_1^{m-1}) \right\},
\end{aligned} \tag{2.23}
$$

where $\mathcal{T}_W$ denotes a set of possible word ending frame sequences for W, and $T \in \mathcal{T}_W$ is a time frame sequence that corresponds to $t_1, \ldots, t_{M_W}$. Consequently, the Viterbi score for a word sequence can be obtained by accumulating the word-level Viterbi scores with their language probabilities.

As mentioned in Section 2.4, a Viterbi score for a single word is obtained as in Fig. 2.6 using Eqs. (2.10) and (2.11). To find $\hat{W}$ efficiently without the enumeration of all Ws, the Viterbi algorithm for a single word is extended by introducing inter-word state transitions between different word HMMs. The possible inter-word transitions are defined by a language model such as an FSG. The n-gram language model can also be used in the same manner because the n-gram model can be viewed as a probabilistic FSG (PFSG) in which a probability is appended to each state transition.

As well as the Viterbi algorithm for a single HMM, it iterates the maximum score selection at every HMM state over T frames. The total number of HMM states becomes $|\bar{S}| \times |\mathcal{Q}|$, where $|\bar{S}|$ is the number of states per word HMM and $|\mathcal{Q}|$ is the number of grammar states. Thus, we no longer have to enumerate all Ws and compute $p(O|W)P(W)$ for each W.

A PFSG is defined with a 7-tuple

$$G = (\mathcal{Q}, \mathcal{V}, \mathcal{E}, \mathcal{I}, \mathcal{F}, P, \pi) \tag{2.24}$$

where

1. $\mathcal{Q}$ is a set of states;

2. $\mathcal{V}$ is a set of word labels, i.e., vocabulary;

3. $\mathcal{E} \subseteq \mathcal{Q} \times \mathcal{Q}$ is a set of state transitions;

4. $\mathcal{I} \subseteq \mathcal{Q}$ is a set of initial states;

5. $\mathcal{F} \subseteq \mathcal{Q}$ is a set of final states;

6. $P : \mathcal{Q} \times \mathcal{Q} \rightarrow [0, 1]$ is a state transition probability function;

7. $\pi : \mathcal{I} \rightarrow [0, 1]$ is an initial state probability function;

If G is a bigram language model, each state is assigned to a word in the vocabulary. Suppose a state p_w indicates the state for the word w. The initial probability $\pi(p_w)$ is the same as the unigram probability $P(w)$. The state transition probability $P(p_w|p_v)$ is the same as the bigram probability $P(w|v)$. With a trigram language model, each state is assigned to a word pair and the state transition probability is set at the trigram probability.

Now we show the Viterbi algorithm for continuous speech recognition, given a grammar G and an input feature vector sequence $O = o_1 \dots o_T$. In this algorithm, each state of G has a word HMM that corresponds to the word label of the state, where we may synthesize each word HMM using subword HMMs as in Fig. 2.5.

When using the word HMMs in the algorithm, each word model is extended to have single initial and final states. Let a word HMM in a grammar state p be $\theta_p = (\mathcal{S}_p, \mathcal{Y}_p, \mathcal{A}_p, \mathcal{B}_p, \Pi_p, \mathcal{F})$, where $\mathcal{A}_p = \{a_{\sigma s}^{(p)}|\sigma, s \in \mathcal{S}_p\}$, $\mathcal{B}_p = \{b_s^{(p)}(o)|s \in \mathcal{S}_p, o \in \mathcal{Y}_p\}$, and $\Pi_p = \{\pi_s^{(p)}|s \in \mathcal{S}_p\}$. We denote the initial and final states by i_p and f_p, where the word HMM is assigned to a grammar state p. We also introduce state transition probabilities $a_{i_p s}$ and $a_{s f_p}$ for an arbitrary HMM state $s \in \mathcal{S}_p$, where $a_{i_p s}$ is set at the initial state probability $\pi_s^{(p)}$, and $a_{s f_p}$ is set at a transition probability of leaving the word HMM via state s. No word model is assigned to a null state but we assume it has a single HMM state such that $i_p = f_p$.

When we use a grammar in Fig. 2.8, the whole search network for decoding is organized as in Fig. 2.9 by embedding each word HMM into a corresponding grammar state. The word HMM

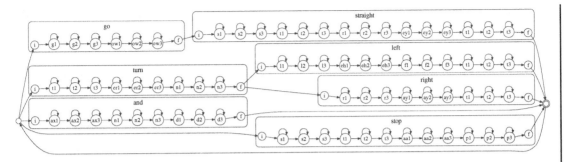

Figure 2.9: A search network for decoding

is synthesized for the grammar state based on the pronunciation of the word, and corresponding subword HMMs, single initial and final states are also connected at the entry and exit of the grammar state.

Given a decoding graph, the most likely word sequence for an utterance can be obtained using a one-pass Viterbi algorithm. This algorithm is also known as a *time-synchronous Viterbi search*, and is based on the time-synchronous computations designed to obtain the following quantities.

$\tilde{\alpha}(t, p, s)$: The Viterbi score of a partial path up to time frame t, at an HMM state s in a grammar state p.

$B(t, p, s)$: A back pointer to keep track of the most likely word sequence up to time frame t at an HMM state s in a grammar state p. $B(t, p, s)$ takes a pair $\langle \tau, q \rangle$, where τ indicates the starting frame of the word assigned to grammar state p, and q is the most likely grammar state right before p. Let $q = 0$ if there is no grammar state before p.

At the ending frame of the utterance, the most likely word sequence can be found by back-tracking the preceding grammar states using back pointers from the best final state with the highest Viterbi score.

We show the basic steps of the algorithm below. Here, let $\mathrm{Adj}(s)$ be the adjacency list of state s and $\mathrm{Word}(p)$ be the word label of grammar state p. If p is a null state, $\mathrm{Word}(p)$ returns ε, which stands for the null string.

Step 1: Initialization
    For each grammar state $p \in \mathcal{Q}$,
      For each HMM state $s \in \mathcal{S}_p$,

$$\tilde{\alpha}(0, p, s) = \begin{cases} \pi(p) & \text{if } p \in \mathcal{I} \text{ and } s = i_p \\ \max_{q \in \mathcal{Q}} \tilde{\alpha}(0, q, f_q) P(p|q) & \text{if } p \notin \mathcal{I} \text{ and } s = i_p \\ 0 & \text{otherwise} \end{cases} \tag{2.25}$$

$$B(0, p, i_p) = \langle 0, 0 \rangle \tag{2.26}$$

Step 2: Time-synchronous processing
For time frames: $t = 1, 2, \ldots, T$,

- Intra-word transition
  For each grammar state $p \in \mathcal{Q}$,
    For each HMM state $s \in (\mathcal{S}_p - \{i_p, f_p\})$,

$$\tilde{\alpha}(t, p, s) = \max_{\sigma \in (\mathcal{S}_p - \{f_p\})} \tilde{\alpha}(t - 1, q, \sigma) a_{\sigma s}^{(p)} b_s^{(p)}(o_t) \tag{2.27}$$

$$B(t, p, s) = B(t - 1, p, \sigma^{max}) \tag{2.28}$$

  For the word-final HMM state f_p,

$$\tilde{\alpha}(t, p, f_p) = \max_{s \in (\mathcal{S} - \{i_p, f_p\})} \tilde{\alpha}(t, q, s) a_{s f_q} \tag{2.29}$$

$$B(t, p, f_p) = B(t, p, s^{max}) \tag{2.30}$$

- Inter-word transition
  For each grammar state $p \in \mathcal{Q}$,

$$\tilde{\alpha}(t, p, i_p) = \max_{q \in \mathcal{Q}} \tilde{\alpha}(t, q, f_q) P(p|q) \tag{2.31}$$

$$B(t, p, i_p) = \langle t, q^{max} \rangle \tag{2.32}$$

Step 3: Termination

$$\hat{\alpha} = \max_{p \in \mathcal{F}} \tilde{\alpha}(T, p, f_p) \tag{2.33}$$

$$\hat{B} = \langle T, p^{max} \rangle \tag{2.34}$$

Step 4: Backtracking

$$\hat{W} = \varepsilon \tag{2.35}$$

$$\langle \hat{t}, \hat{p} \rangle = \hat{B} \tag{2.36}$$

$$\text{while } \langle \hat{t}, \hat{p} \rangle \neq \langle 0, 0 \rangle$$

$$\hat{W} = \text{Word}(\hat{p}) \cdot \hat{W} \tag{2.37}$$

$$\langle \hat{t}, \hat{p} \rangle = B(\hat{t}, \hat{p}, f_{\hat{p}}) \tag{2.38}$$

In Step 1, Viterbi scores and back pointers are initialized with Eqs. (2.25) and (2.26), respectively. The Viterbi score for an initial state i_p of each word HMM assigned to one of the initial grammar states $p \in \mathcal{I}$ is set by the initial probabilities $\pi(p)$. The Viterbi scores for null states that can be visited from one of the initial grammar states are also calculated in Eq. (2.25), where we assume that the grammar state p is referred to in the topological order, i.e., when we calculate $\tilde{\alpha}(0, p, i_p)$, $\tilde{\alpha}(0, q, i_q)$ is already obtained for every state q that can transit to state p. There must be no cycles with only null states in the grammar. Scores for all the other HMM states are set at 0. All back pointers are initially set at $\langle 0, 0 \rangle$.

In Step 2, the Viterbi scores and the back pointers are updated time-synchronously. At each time frame t, intra-word and inter-word transitions are performed. In the intra-word transition, the Viterbi score and the back pointer are obtained with Eqs. (2.28) and (2.28), respectively. In Eq. (2.28), s^{max} indicates the best preceding state in $\mathrm{Adj}(\sigma)$ that gives the maximum score in Eq. (2.28). Thus, the back pointer is simply copied from that of the best preceding state in the intra-word transitions. Then, the scores and pointers are also calculated for the final state f_p in each word HMM using Eqs. (2.30) and (2.30), respectively.

In the inter-word transition, the maximum score and the back pointer are delivered from the final state of each word HMM to the initial state of another word HMM, which can be allowed by the grammar. In Eq. (2.32), the Viterbi score for the best preceding grammar state is selected. In Eq. (2.32), the back pointers are renewed and set at those of the initial states of the next word HMMs, where the current time frame t as the ending frame of the previous word, the best preceding grammar state p^{max}, and the back pointer of the best preceding state $B(t, q^{max}, f_{q^{max}})$ are generated. Transitions for null states in the grammar are also handled in this step, but they need to be processed according to the topological order for the null states.

Finally, Step 3 finds the best score and the back pointer in the final states according to Eqs. (2.34) and (2.34), respectively. Step 4 performs backtracking, which traces back pointers from the best-scored final state and obtains the most likely word sequence. The word sequence $\hat{W}$ is derived using Eqs. (2.37) to (2.38). First $\hat{W}$ is set at ε that satisfies $W \cdot \varepsilon = \varepsilon \cdot W = W$, where "$\cdot$" represents a binary operation that concatenates two strings. Using Eqs. (2.37) to (2.38), $\hat{W}$ is constructed from the end of the sentence by tracking the back pointer $\langle \hat{t}, \hat{p} \rangle$ until $\langle \hat{t}, \hat{p} \rangle = \langle 0, 0 \rangle$, i.e., the beginning of the sentence hypothesis.

2.7.2 TIME-SYNCHRONOUS VITERBI BEAM SEARCH

The basic algorithm described in the previous section does not require the enumeration of all possible word sequences. However, it does require the computation of $O(|\bar{\mathcal{S}}|^2 \times |\mathcal{Q}|^2 \times T)$. Since $|\mathcal{Q}|$ is very large in LVCSR, further reduction of the required computation is crucial. The most widely used technique for this purpose is *beam search* [Low76, NHUTO92]. With this technique, the maximum Viterbi score at each time frame

$$L_t^{max} = \max_{p \in \mathcal{Q}, s \in \mathcal{S}_p} \tilde{\alpha}(t, p, s), \tag{2.39}$$

Algorithm 1 TimeSynchronousViterbiBeamSearch(G, Θ, O)

1: $\langle N, H \rangle \leftarrow$ initialize(G)
2: $\langle N, H_F \rangle \leftarrow$ interword_transition($G, N, H, 0$)
3: **for** $t \leftarrow 1$ to T **do**
4:     $\langle N, H \rangle \leftarrow$ intraword_transition(G, Θ, N, o_t, t)
5:     $\langle N, H_F \rangle \leftarrow$ interword_transition(G, N, H, t)
6:     prune(N, t)
7: **end for**
8: $\hat{B} \leftarrow$ terminate(G, H_F, T)
9: $\hat{W} \leftarrow$ backtrack($\hat{B}$)
10: **return** $\hat{W}$

---

is used to prune unpromising partial paths, i.e., if $\tilde{\alpha}(t, p, s)$ is smaller than γL_t^{max}, the decoder marks the state *inactive* at this frame and does not make a transition from the inactive state at the next frame, where γ is assumed to be a constant such that $0 < \gamma < 1$, which is used to control the degree of beam pruning. Since the scores are accumulated in the log domain, $(\log L_t^{max} - \eta)$ is used as the threshold for pruning, where $\eta = -\log \gamma$. η is often called *beam width*. We can also limit the number of active states by a predetermined number K, i.e., at most K best states are kept active at each frame. K is also called *beam width*. This number-based pruning can be used together with the above score-based pruning. Thus, we can eliminate the computation for the pruned paths, but we need to choose appropriate η and K so that we do not miss the correct path as a result of pruning.

If we incorporate the beam search in the Viterbi algorithm, it is important to handle active/inactive states efficiently. We present an example of a more practical version of the one-pass Viterbi algorithm including beam pruning, which is called a *time-synchronous Viterbi beam search*. Algorithm 1 shows its pseudo code, which consists of the component Algorithms 2 to 7.

Algorithm 1 is basically the same as the procedure we presented in Eqs. 2.25 to 2.34. But active/inactive states are handled efficiently using *queues*. N is a queue of active state pairs in the grammar and the HMM. At line 1, initial state pairs are inserted in N using Algorithm 2. H is a queue of active grammar states that should be taken into account in the inter-word transitions. H_F is a queue of active final grammar states. The queue discipline, i.e., the rule that determines the order of the components to be popped from the queue, is not defined for N, H, and H_F, i.e., any order is allowed. But if a topological order is used for H, the computation for null state transitions in the grammar can be minimized.

In line 2, inter-word transitions are made for initial transitions among null states. Lines 3 to 7 iterate the state transitions at each frame, which consist of intra-word and inter-word state transition steps and a pruning step. The most likely word sequence $\hat{W}$ is obtained by backtracking at line 8, and returned at line 9. Next we briefly explain these component algorithms.

Algorithm 2 performs initialization for the Viterbi search. In line 1, queues N and H are initialized as empty queues. In lines 3 and 4, Viterbi scores and back pointers for initial grammar

---

Algorithm 2 initialize(G)

1:   $N \leftarrow H \leftarrow \emptyset$
2:   **for** each $p \in \mathcal{I}$ **do**
3:      $\tilde{\alpha}(0, p, i_p) \leftarrow \pi(p)$
4:      $B(0, p, i_p) \leftarrow$ null
5:      **if** Word(p) $\neq \varepsilon$ **then**
6:         Enqueue($N, \langle p, i_p \rangle$)
7:      **else**
8:         Enqueue(H, p)
9:      **end if**
10:   **end for**
11:   **return** $\langle N, H \rangle$

---

states and initial HMM states at frame 0 are set at their initial probabilities and the null pointer, respectively. Each pair consisting of initial grammar state p and its initial HMM state i_p is inserted in queue N at line 5. In lines 6-8, only when initial grammar state p is a null state, i.e., Word(p) $= \varepsilon$, p is inserted in H for null transitions from state p.

Algorithm 3 performs intra-word transitions at each frame. In line 1, queues N' and H are initialized. In lines 2-15, state transitions in each word HMM are made from each active HMM state in N, and new active state pairs are inserted in N' for the next frame. In lines 16-27, state transitions to the final state of the word HMM are made from active states in N', and the grammar state that has an active final state in the word HMM is inserted in H. This means that the grammar state is ready for transition.

Algorithm 4 performs inter-word transitions, i.e., grammar level state transitions are made. Each grammar state in H that is ready for transition is picked up, and its score and back pointer are propagated from the final state of the word HMM to the initial state of another word HMM of a succeeding grammar state. Line 9 checks whether or not the grammar state is a null state. If the state is null, it is inserted in H for further transitions. If it is not null, it is inserted in N for intra-word transitions at the next frame.

Algorithm 5 prunes unpromising active states. First, the maximum Viterbi score, L_t^{max}, is obtained at the current frame t. Active states whose score is less than γL_t^{max} are removed from queue N.

Algorithm 6 obtains the recognition result. In lines 3-9, the final state with the best score is found in H_F. Algorithm 7 backtracks the back pointers from the best final state. In line 1, $\hat{W}$ is initialized. In lines 2-6, the most likely word sequence is traced backward from $\hat{B}$, and $\hat{W}$ is constructed.

Algorithm 3 intraword_transition(G, Θ, N, o_t, t)

1:   $N' \leftarrow H \leftarrow \emptyset$
2: **while** $N \neq \emptyset$ **do**
3:     $\langle p, \sigma \rangle \leftarrow \text{Head}(N)$
4:     $\text{Dequeue}(N)$
5:     **for** each $s \in \text{Adj}(\sigma)$ such that $s \neq f_p$ **do**
6:       $l_s = \tilde{\alpha}(t-1, p, \sigma)a_{\sigma s}^{(p)}b_s^{(p)}(o_t)$
7:       **if** $\tilde{\alpha}(t, p, s) < l_s$ **then**
8:         $\tilde{\alpha}(t, p, s) \leftarrow l_s$
9:         $B(t, p, s) \leftarrow B(t-1, p, \sigma)$
10:         **if** $\langle p, s \rangle \notin N'$ **then**
11:           $\text{Enqueue}(N', \langle p, s \rangle)$
12:         **end if**
13:       **end if**
14:     **end for**
15: **end while**
16: **for** each $\langle p, s \rangle \in N'$ **do**
17:     **if** $f_p \in \text{Adj}(s)$ **then**
18:       $l_f = \tilde{\alpha}(t, p, s)a_{sf_p}^{(p)}$
19:       **if** $\tilde{\alpha}(t, p, f_p) < l_f$ **then**
20:         $\tilde{\alpha}(t, p, f_p) \leftarrow l_f$
21:         $B(t, p, f_p) \leftarrow B(t, p, s)$
22:         **if** $p \notin H$ **then**
23:           $\text{Enqueue}(H, p)$
24:         **end if**
25:       **end if**
26:     **end if**
27: **end for**
28: **return** $\langle N', H \rangle$

2.7.3 PRACTICAL TECHNIQUES FOR LVCSR

The time-synchronous Viterbi beam search described in the previous section is a general algorithm for continuous speech recognition. However, when it is applied to LVCSR, we face crucial problems in terms of computation amount and search error. Here we describe the problems and introduce practical solutions used for traditional LVCSR systems.

An n-gram language model is used as a grammar in most LVCSR systems. The n-gram model basically allows all possible concatenations between words. Thus, the number of transitions outgoing from one grammar state is equal to the vocabulary size, e.g., 65 thousand. This means that

Algorithm 4 interword_transition(G, N, H, t)

1: $H_F \leftarrow \emptyset$
2: **while** $H \neq \emptyset$ **do**
3:    $q \leftarrow \text{Head}(H)$
4:    $\text{Dequeue}(H)$
5:    **for** each $p \in \text{Adj}(q)$ **do**
6:      $l_i \leftarrow \tilde{\alpha}(t, q, f_q) P(p|q)$
7:      **if** $\tilde{\alpha}(t, p, i_p) < l_i$ **then**
8:        $\tilde{\alpha}(t, p, i_p) \leftarrow l_i$
9:        $B(t, p, i_p) \leftarrow \langle t, q \rangle$
10:       **if** $\text{Word}(p) \neq \varepsilon$ **then**
11:         **if** $\langle p, i_p \rangle \notin N$ **then**
12:           $\text{Enqueue}(N, \langle p, i_p \rangle)$
13:         **end if**
14:       **else**
15:         **if** $p \notin H$ **then**
16:           $\text{Enqueue}(H, p)$
17:         **end if**
18:       **end if**
19:      **end if**
20:    **end for**
21:    **if** $q \in \mathcal{F}$ and $q \notin H_F$ **then**
22:      $\text{Enqueue}(H_F, q)$
23:    **end if**
24: **end while**
25: **return** $\langle N, H_F \rangle$

the number of active states is greatly increased by the transitions from the final state of each word HMM to all the initial states of the succeeding word HMMs. In this case, it is difficult to reduce the active states safely by beam pruning, because there are many active states with equivalent Viterbi scores in the initial parts of the word HMMs.

One solution is to use a pronunciation prefix tree. Since there are many words that have the same pronunciation prefix in a large vocabulary, the number of outgoing transitions can be decreased by sharing such prefixes in the search network. Suppose a vocabulary consists of the words *start*, *stop*, *straight*, and *go*. The pronunciation prefix tree can be constructed as in Fig. 2.10. In the search network, a phone HMM is embedded in each phone node and the leaf nodes of the tree are linked to the root node. Each node of a rectangle is a final HMM state for identifying which word corresponds to the path from the root node to the leaf node. By introducing this tree structure in the search network, the number of outgoing transitions from each HMM state becomes at most the

---

Algorithm 5 prune(N, t)

1: $L_t^{max} \leftarrow \max\limits_{\langle p,s \rangle \in N} \tilde{\alpha}(t, p, s)$
2: **for** each $\langle p, s \rangle \in N$ **do**
3:   **if** $\tilde{\alpha}(t, p, s) < \gamma L_t^{max}$ **then**
4:     Erase($N, \langle p, s \rangle$)
5:   **end if**
6: **end for**
7: **return** N

Algorithm 6 terminate(G, H_F, T)

1: $\hat{\alpha} \leftarrow 0$
2: $\hat{B} \leftarrow \langle 0, 0 \rangle$
3: **for** each $q \in H_F$ **do**
4:   $l_f = \tilde{\alpha}(T, q, f_q)$
5:   **if** $\hat{\alpha} < l_f$ **then**
6:     $\hat{\alpha} \leftarrow l_f$
7:     $\hat{B} \leftarrow \langle T, q \rangle$
8:   **end if**
9: **end for**
10: **return** $\hat{B}$

Algorithm 7 backtrack($\hat{B}$)

1: $\hat{W} \leftarrow \varepsilon$
2: $\langle \hat{t}, \hat{p} \rangle \leftarrow \hat{B}$
3: **while** $\langle \hat{t}, \hat{p} \rangle \neq \langle 0, 0 \rangle$ **do**
4:   $\hat{W} \leftarrow \text{Word}(\hat{p}) \cdot \hat{W}$
5:   $\langle \hat{t}, \hat{p} \rangle \leftarrow B(\hat{t}, \hat{p}, f_{\hat{p}})$
6: **end while**
7: **return** $\hat{W}$

---

number of phones, which is much less than the number of words in LVCSR. Thus, the pronunciation prefix tree is very important as regards avoiding irruptions of active states in the beam search.

However, when we use a prefix tree, we need to consider that the word identity is unknown until the path reaches a leaf node. Since grammar-level state transitions cannot be made until the word identity is determined, we need to associate the tree with each grammar state. In the case of n-gram language models, $|V|^{n-1}$ trees are necessary, and each of them depends on the history of $n - 1$ words. Figure 2.11(a) shows a search network with prefix trees based on a bigram language model of a vocabulary $\{u, v, w\}$. The tree on the left-hand side is the unigram tree whose root node

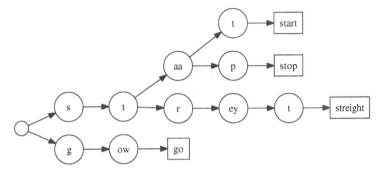

Figure 2.10: Pronunciation prefix tree

becomes active first in the Viterbi search. Then active states are propagated to the leaf nodes and each word is identified if its leaf node is activated. After that, a transition is made to a successor tree, where the tree is conditioned by the $n - 1$-word history of the path, i.e., the previously identified word in the bigram case.

Although the total number of transitions becomes much larger than that without the prefix trees, the tree organization is very effective in the beam search. However, a large memory is needed to store an entire search network. This typically requires the incremental construction of the network during the search to save memory, which generates a substantial computation overhead. To reduce the number of transitions, there is a technique for introducing back-off transitions into the network [ABCF95, BC95]. If there is no bigram entry for a word at a leaf node conditioned by the previous word, a transition to the unigram tree is made with the back-off weight. Using this technique, the history-conditioned tree includes only existing n-gram entries. Figure 2.11(b) shows an example where there are no bigram entries for $P(w|v)$, $P(u|w)$, and $P(w|w)$ in the model. The network is represented as a small number of transitions compared with that in Fig. 2.11(a). However, with this method, a back-off transition from an n-gram to its corresponding $n - 1$-gram can be made even though the n-gram entry exists in the model. For example, in Fig. 2.11(b), if $P(u|v) < \alpha(v)P(u)$, the path going through $\alpha(v)P(u)$ is selected even if the entry $P(u|v)$ exists. Since $P(u|v) > \alpha(v)P(u)$ is usually satisfied, it has little impact on the recognition results, but the score may include a slight error.

Indeed, Algorithms 3 and 4 need to be extended if they are to use the prefix-tree-based search network, because each grammar state has a prefix tree but does not have a single word HMM. The tree has multiple final HMM states each of which corresponds to a word. In this case, we have to consider transitions from each final state to the corresponding initial (root) HMM state of a successor tree in the inter-word transition procedure. In the initial state, the maximum score needs to be selected. The prefix-tree-based search algorithm is not described in detail in this book. The details can be found, for example, in [ONA97].

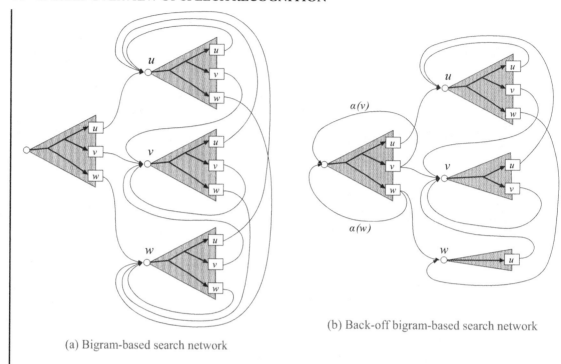

(a) Bigram-based search network

(b) Back-off bigram-based search network

Figure 2.11: Bigram-based search networks with prefix trees

We also need to consider the fact that the n-gram probability cannot be applied until the leaf node. This increases the risk of pruning error. In general, it is better to apply the language scores as soon as possible for the beam search, because the decoder can find unpromising paths earlier and exclude them as candidates. Since it is impossible to apply the correct score before the word identity is determined, a look-ahead score is used instead. The look-ahead score is, for example, calculated at each node as the maximum of the n-gram probabilities of the words that can be reached from that node. This is realized by attaching a factored language score to each transition between phone nodes. The factored score for a transition from node i to node j in the prefix tree of history h is calculated as

$$P(j|i, h) = \frac{\max_{w \in \Omega(j)} P(w|h)}{\max_{v \in \Omega(i)} P(v|h)}, \tag{2.40}$$

where $\Omega(j)$ represents a set of words that can be reached from node j. The accumulation of factored scores along a path from a root node to a leaf node is equal to the language score corresponding to the leaf node.

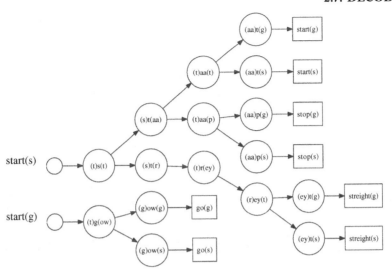

Figure 2.12: Triphone-based pronunciation prefix tree conditioned by preceding word *start*. "start(g)" and "start(s)" represent word labels of *start*, which depend on the succeeding phones /g/ and /s/, respectively.

2.7.4   CONTEXT-DEPENDENT PHONE SEARCH NETWORK

As mentioned in Section 2.5, context-dependent phone models need to be connected consistently with their context information when synthesizing a word model. The pronunciation prefix trees must also be constructed so that each phone node is context-dependent and consistently connected with each other. Figure 2.12 shows an example of a triphone-based prefix tree conditioned by the word *start*. Each node has a triphone that depends on the preceding and succeeding phones. As shown in the figure, the root and leaf nodes are also dependent on the context, i.e., the ending phone of the preceding word and the beginning phone of the succeeding word. At each inter-word transition, the phone context dependency needs to be correctly dealt with in the decoder.

2.7.5   LATTICE GENERATION AND N-BEST SEARCH

In the previous sections, we described the Viterbi search for finding the most likely word sequence. However, multiple hypotheses are often needed by applications using speech recognition. Since a speech signal is essentially ambiguous and difficult to resolve using only acoustic and language scores, it is impossible to avoid recognition errors. Therefore, if the speech recognizer can output multiple hypotheses with high scores, the application can select a reasonable hypothesis from among them using some additional knowledge such as the current status of the application or the topic of the conversation.

On the other hand, multiple hypotheses are also used in some important speech recognition techniques such as multi-pass decoding, discriminative training, unsupervised adaptation, and con-

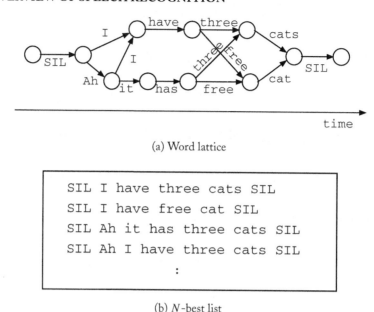

(a) Word lattice

```
SIL I have three cats SIL
SIL I have free cat SIL
SIL Ah it has three cats SIL
SIL Ah I have three cats SIL
              :
```

(b) N-best list

Figure 2.13: Multiple hypotheses represented by (a) word lattice and (b) N-best list.

fidence estimation. In multi-pass decoding, a set of hypotheses is first generated and then rescored with more complex models or adapted models. After rescoring, the final result is selected from among them. This rescoring step can be repeated to obtain a better result. In discriminative training, the multiple hypotheses for training data are used to calculate a loss function, which indicates the risk of occurrence of recognition errors. The acoustic/language models are trained to minimize the loss function. Basically, the model parameters are updated so that the scores for correct hypotheses increase and those for incorrect hypotheses decrease. In unsupervised adaptation, model parameters are estimated using the multiple hypotheses for the adaptation data as a pseudo transcript. It is known that using multiple hypotheses is more effective in mitigating the impact of errors on the adaptation process than using only the best hypothesis. In confidence estimation, the likelihood ratio of the best hypothesis to the other hypotheses is considered a confidence measure of the best hypothesis. Thus, generating multiple hypotheses is currently one of the most important functions of speech recognition engines.

A set of multiple hypotheses is usually represented as a word lattice or an N-best list. An N-best list is a list of N-most likely word sequences sorted by their scores. The word lattice is a directed acyclic graph in which each arc is labeled with a hypothesized word and each node corresponds to a word boundary with the time information. A path from an initial node to a final node in the graph indicates a word sequence, i.e., the graph represents a set of multiple hypotheses. Figure 2.13 shows (a) a word lattice and (b) an N-best list. A word lattice is generally more effective in representing various hypotheses compactly. Each lattice arc has a word label and its score. Each lattice node depends on a grammar state and a time frame. On the other hand, the N-best list is easier to use for post processing such as parsing that analyzes the meaning of each word sequence, because a word sequence like a sentence is assumed to be given as an input in most language processing techniques.

The word lattice can be constructed in a time-synchronous Viterbi beam search. The N-best list can be extracted from the word lattice using the A^* search. First we briefly explain how to generate a word lattice.

A word lattice is a subset of all possible word sequences accepted by the grammar that survived in the beam search. Hence, the lattice size changes depending on the beam width. Basically, the nodes and arcs of a word lattice can be generated at each inter-word state transition by keeping all the back pointers from active states. This means that we do not assign only the best preceding back pointer, but also a set of possible back pointers, i.e., Eq. (2.32) is replaced with

$$B(t, p, i_p) = \{\langle t, q \rangle | q \in \mathcal{Q}\}. \tag{2.41}$$

In the termination step, Eq. (2.34) is also replaced with

$$\hat{B} = \{\langle t, p \rangle | p \in \mathcal{F}\}. \tag{2.42}$$

In Algorithm 4 for inter-word transition, line 9 is changed to

$$B(t, p, i_p) \leftarrow B(t, p, i_p) \cup \{\langle t, q \rangle\}, \tag{2.43}$$

and moved between lines 6 and 7 to keep all active back pointers other than the best pointer. In Algorithm 6 for termination, line 7 is also changed to

$$\hat{B} \leftarrow \hat{B} \cup \{\langle T, p \rangle\}. \tag{2.44}$$

and moved between lines 4 and 5.

The lattice can be constructed with Algorithm 8 in which the nodes and arcs of the lattice are created by tracking the back pointers from active final states in the same manner as the backtracking. But we need to consider multiple hypotheses. In the algorithm, an empty lattice $\mathcal{L}$ is first created in line 1. The active back pointers are inserted into queue H at line 2. An empty queue C is prepared in line 3, which is used to memorize already-created lattice nodes. Note that a lattice node is defined as a pair consisting of a time index and a grammar state. Lines 4 to 16 are repeated until the lattice is completed. In lines 5 and 6, a back pointer is popped from H. Each lattice arc from the pointer is created with function AddArc at line 9. Arc score l calculated in line 8 is also assigned to the arc, where we assume $\tilde{\alpha}(\tau, q, f_q) = 1$ if $q = 0$. Each preceding back pointer is inserted into H in line 13 to make nodes and arcs further back toward the initial state. In line 14, $\langle t, p \rangle$ is inserted in C to memorize a lattice node corresponding to this pair.

Although the above procedure for lattice generation is widely used, we need to consider that it includes an approximation. Since the algorithm is based on the Viterbi search designed to find

---

Algorithm 8 lattice_generation($\hat{B}$)

  1: $\mathcal{L} \leftarrow \emptyset$
  2: $B \leftarrow \hat{B}$
  3: $C \leftarrow \emptyset$
  4: **while** $H \neq \emptyset$ **do**
  5:      $\langle t, p \rangle \leftarrow \text{Head}(B)$
  6:      $\text{Dequeue}(B)$
  7:      **for** each $\langle \tau, q \rangle \in B(t, p, f_p)$ **do**
  8:          $l \leftarrow \tilde{\alpha}(t, p, f_p)/\tilde{\alpha}(\tau, q, f_q)$
  9:          $\text{AddArc}(\mathcal{L}, \langle\langle \tau, q \rangle, \langle t, p \rangle, \text{Word}(p), l\rangle)$
 10:          **if** $\langle \tau, q \rangle \notin C$ **then**
 11:              $\text{Enqueue}(B, \langle \tau, q \rangle)$
 12:          **end if**
 13:      **end for**
 14:      $\text{Enqueue}(C, \langle t, p \rangle)$
 15: **end while**
 16: **return** $\mathcal{L}$

---

only the best path, it cannot be guaranteed that candidates other than the best path are obtained exactly.

We show an example in which an approximation error occurs. Figure 2.14 (a) represents a trellis space, where three paths from grammar states u, v, and x go into the next grammar state w. Suppose the thick line is the best path. The inter-word transitions from u and v to w at time t_1 are retained by making back pointers. In this case, the paths going through uw and vw will be created correctly in the lattice. On the other hand, a path going through an inter-word transition from x to w at time t_2 will be lost at time t_3 because only one score and one back pointer are held at a certain time frame within a grammar state. Thus, even if the path is the second or third best path, it will be lost when it encounters the best path in the grammar state. It is possible to save such a path by making all back pointers in intra-word transitions. However, it much increases memory consumption and computation amount.

There is another approach for efficiently reducing such approximation errors. This is realized by keeping scores and back pointers differently depending on the preceding word, where it is assumed that the best starting time of a word depends only on the preceding word. Although it may depend on earlier preceding words, this assumption, called *word-pair approximation* [ONA97], is much safer than the simple lattice generation method above. Figure 2.14 (b) represents paths in the trellis space depending on the previous word. In this example, all the paths safely reach the end of the grammar state w at time t_4 because the scores and back pointers are differently held within w.

This extension can be made by simply duplicating each grammar state depending on the words attached to the preceding grammar states that have transitions to the original state. Although this

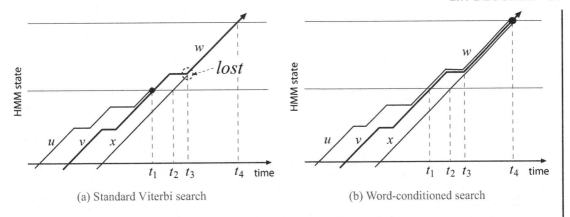

(a) Standard Viterbi search   (b) Word-conditioned search

Figure 2.14: Paths in (a) standard search and (b) word-conditioned search.

duplication seems to increase the amount of computation needed for the Viterbi search, there are many search networks that inherently satisfy this assumption. Actually, a prefix tree-based search network as in 2.11(a) has a grammar state for each tree that depends on the previous word. But 2.11(b) has a word independent tree for back-off transitions. Therefore, a search network such as that in Fig. 2.11(a) is usually used to generate word lattices.

Next we show how to obtain the N-best list by using an A^* search, which is performed backward from the final states. The algorithm is similar to that for lattice generation, and therefore it includes the same approximation [SA90]. Algorithm 9 performs the N-best search in which a priority queue according to the product of the forward and backward Viterbi scores is used to obtain the top N-best word sequences. In the first line, an empty list $\mathcal{W}$ for word sequences is prepared. Priority queue U is initialized in line 2 and elements corresponding to active final states are inserted in U, where each element consists of time, grammar state, word sequence, forward score, and backward score. In lines 6-17, word sequences with highest scores are searched until the size of $\mathcal{W}$ becomes N. In lines 7 and 9, the element is popped from U such that the product of forward and backward scores, $(\alpha \times \beta)$, is the maximum in U. If the preceding grammar state p is 0, the sentence is completed. The word sequence W is inserted in $\mathcal{W}$. If W is already included in $\mathcal{W}$, we may abandon it to eliminate duplication. If $p \neq 0$, new elements are generated based on the back pointers and they are inserted in U. In line 13, the word score for the preceding state q is calculated, where we assume $\tilde{\alpha}(\tau, q, f_q) = 1$ if $q = 0$. The score is then used to obtain new α and β. The word sequence is updated by adding Word(q) before W. Finally, the N-best list including N word sequences is obtained. The N-best list can also be obtained from a word lattice using the same algorithm.

Algorithm 9 nbest_search($\hat{B}$, N)

1: $\mathcal{W} \leftarrow \emptyset$
2: $U \leftarrow \emptyset$
3: **for** each $\langle t, p \rangle \in \hat{B}$ **do**
4:     $U \leftarrow U \cup \{\langle t, p, \varepsilon, \tilde{\alpha}(t, p, f_p), 1 \rangle\}$
5: **end for**
6: **while** $|\mathcal{W}| < N$ and $U \neq \emptyset$ **do**
7:     $\langle t, p, W, \alpha, \beta \rangle \leftarrow \text{Head}(U)$
8:     $\text{Dequeue}(U)$
9:     **if** $p = 0$ **then**
10:         $\mathcal{W} \leftarrow \{W\} \cup \mathcal{W}$
11:     **else**
12:         **for** each $\langle \tau, q \rangle \in B(t, p, f_p)$ **do**
13:             $l \leftarrow \tilde{\alpha}(t, p, f_p) / \tilde{\alpha}(\tau, q, f_q)$
14:             $\text{Enqueue}(U, \langle \tau, q, \text{Word}(p) \cdot W, \tilde{\alpha}(\tau, q, f_q), l \times \beta \rangle)$
15:         **end for**
16:     **end if**
17: **end while**
18: **return** $\mathcal{W}$

CHAPTER 3

Introduction to Weighted Finite-State Transducers

We already mentioned in Chapter 1, that the Weighted Finite-State Transducer (WFST) provides an elegant framework [MPR02] for speech recognition decoding. In this chapter, we formally define WFSTs and describe their basic properties based on automata theory. Then we show some important operations defined on WFSTs, which are used for constructing and optimizing a speech recognition network.

3.1 FINITE AUTOMATA

A WFST is a sort of finite automaton (FA). An FA consists of a finite set of states and state transitions, where each transition has at least one label. The most basic FA is a finite-state acceptor (FSA). Given a sequence of input symbols, an FSA returns *"accepted"* or *"not accepted,"* according to whether or not the FSA has a path, i.e., a state transition process, from an initial state to a final state, whose label sequence matches the input symbol sequence.

Figure 3.1 (a) shows an example of an FSA, where the nodes and arcs correspond to its states and state transitions. For example, the FSA accepts a symbol sequence "a,b,c,d" with state transitions 0, 1, 1, 2, 5, but does not accept "a,b,d." Thus, an FSA represents a set of symbol sequences that can be accepted. A symbol sequence is also called a *string*. The FSA in the figure represents a set of acceptable symbol sequences, which corresponds to a regular expression "ab*cd|bcd*e." We assume here that state 0 is the initial state and state 5 is the final state. Unless noted, let an initial state be depicted with a circle of thick line and a final state be depicted with a double circle in this book.

Next we introduce several extensions of FSAs, where we mention only Finite-State Transducers (FSTs), Weighted Finite-State Acceptors (WFSAs), and Weighted Finite-State Transducers (WFSTs). These automata inherit the basic characteristics of FSAs, but they output not only a binary value "accepted/not accepted" but also a symbol sequence, a weight, or both.

An FST has an output label at each transition, i.e., a pair of input and output labels is assigned. Figure 3.1 (b) shows an example of an FST. A label pair in the form of "input-label : output-label" is placed at each transition. By this extension, an FST can describe a set of rules for conversion or *transduction* from one symbol sequence to another. The example FST converts a symbol sequence "a,b,c,d" into another symbol sequence "z,y,x,w."

A WFSA has a weight at each transition, an initial weight in each initial state, and a final weight in each final state. A weight usually indicates the probability or cost of a transition or initial/final

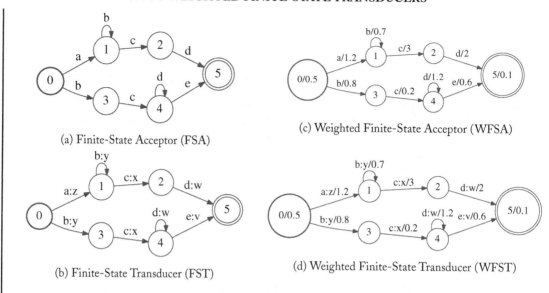

(a) Finite-State Acceptor (FSA)

(c) Weighted Finite-State Acceptor (WFSA)

(b) Finite-State Transducer (FST)

(d) Weighted Finite-State Transducer (WFST)

Figure 3.1: Examples of finite automata

state, and weights are assumed to be cumulated by multiplication along each path and addition over different paths. Thus, the WFSA gives us a measure with which to compare different paths, different sets of paths, or different WFSAs. Figure 3.1 (c) shows an example of a WFSA. As in the figure, the weight at each transition is denoted an "input-label/weight" and weights for initial and final states are denoted "state number/weight," where the initial state 0 possesses an initial weight 0.5 and the final state 5 possesses a final weight 0.1. This WFSA, for example, accepts a sequence "a,b,c,d" with a cumulative weight "0.252" as a result of multiplication $0.5 \times 1.2 \times 0.7 \times 3 \times 2 \times 0.1$ along the path of state transitions 0,1,1,2,5.

A WFST has both an output label and a weight in addition to an input label at each transition, i.e., the WFST inherits the properties of both the FST and WFSA. Figure 3.1 (d) shows an example of a WFST, where the label on each transition represents the "input-label : output-label / weight" of the transition. Initial and final weights are also assigned to the corresponding states. This WFST, for example, converts a symbol sequence "a,b,c,d" into "z,y,x,w" with a cumulative weight "0.252."

A WFST over a set of weight elements $\mathbb{K}$ is formally defined by an 8-tuple $(\Sigma, \Delta, Q, I, F, E, \lambda, \rho)$ where:

1. Σ is a finite set of input labels;

2. Δ is a finite set of output labels;

3. Q is a finite set of states;

4. $I \subseteq Q$ is a set of initial states;

5. $F \subseteq Q$ is a set of final states;

6. $E \subseteq Q \times (\Sigma \cup \{\varepsilon\}) \times (\Delta \cup \{\varepsilon\}) \times \mathbb{K} \times Q$ is a finite multi-set of transitions;

7. $\lambda : I \rightarrow \mathbb{K}$ is an initial weight function;

8. $\rho : F \rightarrow \mathbb{K}$ is a final weight function.

"ε" is a meta symbol label that indicates there is no symbol to input or output. According to the above form, the WFST in Fig. 3.1(d) can be defined as:

1. $\Sigma = \{a,b,c,d,e\}$,

2. $\Delta = \{v,x,y,w,z\}$,

3. $Q = \{0, 1, 2, 3, 4, 5\}$,

4. $I = \{0\}$,

5. $F = \{5\}$,

6. $E = \{(0,a,z,1.2,1), (0,b,y,0.8,3), (1,b,y,0.7,1), (1,c,x,3,2), (2,d,w,2,5), (3,c,x,0.2,4),$
   $(4,d,w,1.2,4), (4,e,v,0.6,5)\}$,

7. $\lambda(0) = 0.5$,

8. $\rho(5) = 0.1$,

where each transition in E is represented as (source state, input label, output label, weight, destination state). The other FAs, i.e., FSA, FST, and WFSA, can be viewed as special cases of a WFST and can also be defined in similar ways by omitting weights and/or output labels.

There is another extension for representing a finite automaton in which a label of each transition is allowed to be an input string (and an output string) rather than a single input symbol (and a single output symbol). However, in this book we focus on the case where each transition accepts (and outputs) at most one symbol at a transition. Basically, such a string-based automaton can be converted into its equivalent symbol-based automaton by substituting each string-based transition with a chain of symbol-based transitions.

Finally, we summarize the four types of automata in Table 3.1. As mentioned above, an FSA has only an input label on each transition and returns accepted or not accepted for an input symbol sequence. In the table, "function" means the mapping that can be represented by the automaton. With an FSA, an input symbol sequence in Σ^* can be mapped to a binary value in $\{0, 1\}$, where we assume that 0 and 1 correspond to "accepted" and "not accepted," respectively. An FST has input and output labels on each transition and can represent a function $f : \Sigma^* \rightarrow 2^{\Delta^*}$, where 2^{Δ^*} is the power set of Δ^*. This means that an FST can convert an input symbol sequence into a set of output symbol sequences. An WFSA has an input label and a weight on each transition and represents a mapping $f : \Sigma^* \to \mathbb{K}$, i.e., given an input symbol sequence, it returns a weight for the sequence. An WFST represents a mapping to a set of weighted output symbol sequences from a given input symbol sequence.

Table 3.1: Types of finite automaton

type	input	output	weight	mapping
Finite-state Acceptor (FSA)	✓			$\Sigma^* \to \{0, 1\}$
Finite-state Transducer (FST)	✓	✓		$\Sigma^* \to 2^{\Delta^*}$
Weighted FSA (WFSA)	✓		✓	$\Sigma^* \to \mathbb{K}$
Weighted FST (WFST)	✓	✓	✓	$\Sigma^* \to 2^{\Delta^*} \times \mathbb{K}$

3.2 BASIC PROPERTIES OF FINITE AUTOMATA

An important characteristic of a finite automaton is whether the automaton is *deterministic* or *non-deterministic*. A deterministic FA (DFA) has only one single initial state and at most one transition for any input label from each state. This means that the transition made by the FA is unique at a given state for a given symbol, and therefore the destination state is also unique. Accordingly, there is only one path from the initial state to the final state for an input symbol sequence if it is accepted. Thus, DFAs have the advantage of computation with which to obtain an output for an input symbol sequence. Actually, the complexity is $O(L \log_2 \hat{D})$ if we use a binary search to find a transition with an input label matched to every input symbol in a sequence of length L, where $\hat{D}$ denotes the maximum number of transitions outgoing from a state in the automaton. The complexity is linear to the length L but not significantly dominated by $\hat{D}$, i.e., the structure of the automaton does not have a big impact on the computation time.

On the other hand, a non-deterministic FA (NFA) can have more than one transition from a state for an input label. Accordingly, we have to consider multiple paths for an input symbol sequence. Although the computational complexity depends on the structure of the NFA, it becomes $O(L \times |Q| \times |E|)$ in the worst case. However, there is an algorithm for converting an NFA into an DFA that is equivalent to the original NFA. This is well known as the *determinization* algorithm. With determinization, the function of the NFA can be applied to an input sequence with a small amount of computation using its equivalent DFA. Figure 3.2 shows examples of NFA and DFA that have the same function. Note that although all FSAs are determinizable, this is not always the case with other FAs such as FSTs, WFSAs, and WFSTs.

Transducers, i.e., FSTs and WFSTs, are defined as *sequential* if and only if they are deterministic with respect to the input labels of their transitions. Moreover, transducers are defined as *functional* if and only if they have at most one output symbol sequence for any given input symbol sequence. It is known that functional transducers are determinizable. Formal definitions are given in [RS97]. Figure 3.3 shows some types of FSTs. Figure 3.3(a) is an example of functional FST and Fig. 3.3(b) is its sequential FST obtained by determinization. Figure 3.3(c) is an example of non-functional

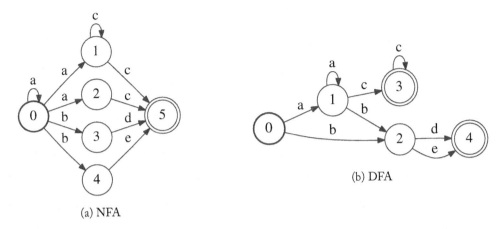

(a) NFA

(b) DFA

Figure 3.2: Non-deterministic and deterministic finite automata

FST. This FST is determinizable as regards input labels, but there is an ambiguity as regards output labels. In this case, its determinization results in a *p-subsequential* FST as in Fig. 3.3(d), which is a more general class of sequential FSTs. A p-subsequential FST is allowed to have p final emission labels at final states, where $p \leq 0$. However, the representation including final emission labels is beyond the definition of a WFST in this book. For such FSTs, we can use a pseudo representation as in Fig. 3.3(e), which is equivalent to the p-subsequential FST in Fig. 3.3(d).

Another important property of finite automata is whether or not they have at least one transition such that the input label is epsilon. Epsilon is the empty string denoted by "ε." Such a transition is called an *epsilon transition*, and the state transition can be made without any input. Figure 3.4 (a) shows an example of an FSA that has epsilon transitions. This FSA first makes a transition from state 0 to 1 for input symbol "a" or "b." Since there is an epsilon transition from state 1 to state 2, the FSA can make a transition to state 3 before reading the next symbol. But it may remain at state 2. This means that the FSA can be in states 2 and 3 simultaneously. Thus, FAs with epsilon transitions are non-deterministic and called ε-NFAs.

There is an operation that can eliminate epsilon transitions from an ε-NFA. The *epsilon removal* operation converts an ε-NFA into one without epsilon transitions, that is equivalent to the original ε-NFA. The FSA in Fig. 3.4 (a) is shown in Fig. 3.4 (b) with epsilon removed.

Finally, we mention a minimal DFA that has the minimum number of states in a set of equivalent DFAs. The *minimization* operation finds the minimal DFA that is equivalent to a given DFA. Figure 3.5 shows an example of (a) a DFA and (b) its minimal DFA. This property is also used to check the equivalence of FAs, since equivalent FAs after epsilon removal, determinization, and minimization have exactly the same set of states and transitions. The properties mentioned above are formally described in textbooks on finite automata [HMU06].

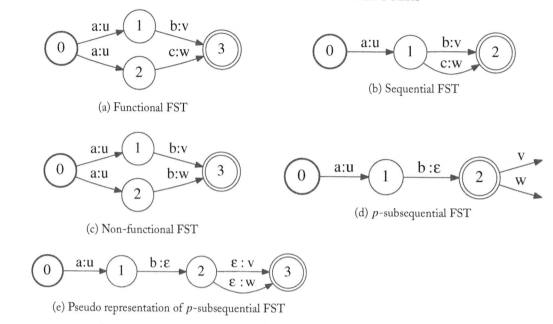

(a) Functional FST

(b) Sequential FST

(c) Non-functional FST

(d) p-subsequential FST

(e) Pseudo representation of p-subsequential FST

Figure 3.3: Finite-state transducers with different properties

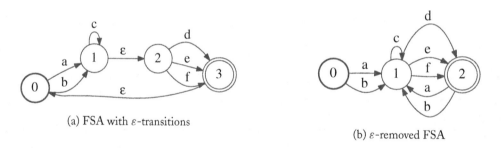

(a) FSA with ε-transitions

(b) ε-removed FSA

Figure 3.4: ε-NFA and ε-free NFA

3.3   SEMIRING

In weighted automata, the weights and their binary operations "addition" and "multiplication" are formally defined to generalize the automata and their algorithms. In the theory, weights and their operations are defined by a *semiring*, which is an algebraic structure in abstract algebra. This means that any kinds of weights can be dealt with in the automata algorithms if a semiring can be defined over the set of weights.

A semiring is similar to a *ring*, which is also an algebraic structure, but the existence of *additive inverse* is not required. A semiring is defined as $(\mathbb{K}, \oplus, \otimes, \bar{0}, \bar{1})$, where $\mathbb{K}$ is a set of elements, $\oplus$ and

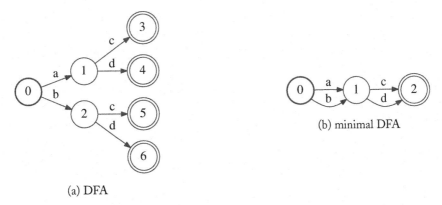

(a) DFA

(b) minimal DFA

Figure 3.5: DFA and its minimal DFA

$\otimes$ are two formally defined binary operations, i.e., "addition" and "multiplication," over $\mathbb{K}$, $\bar{0}$ is an additive identity element, and $\bar{1}$ is a multiplicative identity element. The semiring must satisfy the axioms listed in Table 3.2.

Table 3.2: Algebraic structure of semiring, that satisfies the listed axioms for all $x, y, z \in \mathbb{K}$	
Associativity for addition	$(x \oplus y) \oplus z = x \oplus (y \oplus z)$
Commutativity for addition	$x \oplus y = y \oplus x$
Associativity for mutiplication	$(x \otimes y) \otimes z = x \otimes (y \otimes z)$
Distributivity for multiplication over addition	$(x \oplus y) \otimes z = (x \otimes z) \oplus (y \otimes z)$ $x \otimes (y \oplus z) = (x \otimes y) \oplus (x \otimes z)$
Property of additive identity	$\bar{0} \oplus x = x \oplus \bar{0} = x$ $\bar{0} \otimes x = x \otimes \bar{0} = \bar{0}$
Property of multiplicative identity	$\bar{1} \otimes x = x \otimes \bar{1} = x$

Table 3.3 shows some semirings used in WFST-based applications. $\oplus_{\log}$ denotes a binary

Table 3.3: Semirings					
Semiring	$\mathbb{K}$	$\oplus$	$\otimes$	$\bar{0}$	$\bar{1}$
Probability	$[0, 1]$	$+$	$\times$	0	1
Log	$[-\infty, +\infty]$	$\oplus_{\log}$	$+$	∞	0
Tropical	$[-\infty, +\infty]$	$\min$	$+$	∞	0
String	$\Sigma^* \cup \{s_\infty\}$	$\wedge$	$\cdot$	s_∞	ε

operation $x \oplus_{\log} y = -\log(e^{-x} + e^{-y})$ for any x and y in $\mathbb{K}$. $\wedge$ denotes the longest common prefix between two strings. In WFST-based speech recognition, the tropical semiring is mainly used, which consists of a set of real-valued weights with "addition" and "multiplication" defined as the minimum of the two and ordinary addition, respectively. In some optimization steps for WFSTs, the log semiring is also used.

When we apply an operation to a weighted automaton, it is meaningful to consider certain properties of the semiring we are using, because the algorithm and computation of the operation may be simplified by using the properties. We describe some of these properties below.

- *Commutative*;
  A semiring $(\mathbb{K}, \oplus, \otimes, \bar{0}, \bar{1})$ is commutative if its multiplication is commutative, i.e.,

$$x \otimes y = y \otimes x$$

  for all x and y in $\mathbb{K}$. The tropical and log semirings are commutative.

- *Idempotent*;
  A semiring $(\mathbb{K}, \oplus, \otimes, \bar{0}, \bar{1})$ is idempotent if its addition "$\oplus$" satisfies

$$x = x \oplus x$$

  for all x in $\mathbb{K}$. The tropical semiring is idempotent, i.e.,

$$x = \min(x, x),$$

  but the log semiring is not, i.e.,

$$x \neq x \oplus_{\log} x$$

  for some x in $\mathbb{K}$.

- *k-closed semiring*;
  Let $k \geq 0$ be an integer. A semiring $(\mathbb{K}, \oplus, \otimes, \bar{0}, \bar{1})$ is k-closed if

$$\bigoplus_{n=0}^{k+1} x^n = \bigoplus_{n=0}^{k} x^n$$

  for all x in $\mathbb{K}$. For any integer l such that $l > k$,

$$\bigoplus_{n=0}^{l} x^n = \bigoplus_{n=0}^{k} x^n.$$

can be proved with the mathematical induction. The tropical semiring is obviously 0-closed. The log semiring can be assumed to be k-closed if we ignore very small differences.

- *Weakly left-divisible*;
  A semiring $(\mathbb{K}, \oplus, \otimes, \bar{0}, \bar{1})$ is weakly left-divisible if for any x and y in $\mathbb{K}$ such that $x \oplus y \neq \bar{0}$, there exists at least one z such that $x = (x \oplus y) \otimes z$. The tropical and log semirings are both weakly left-divisible.

- *Zero-sum free*;
  A semiring is zero-sum free if $x \oplus y = \bar{0}$ implies $x = y = \bar{0}$ for any x and y in $\mathbb{K}$.

3.4   BASIC OPERATIONS

Here, we overview unary and binary operations defined over finite automata. We have already stated that a finite automaton represents a set of (weighted) symbol sequences or (weighted) transductions between symbol sequences. The basic operations extend the set by adding or removing transitions and combining it with another automaton.

In automata theory, three rational operations, *Kleene closure*, *union*, and *concatenation* are defined over FAs. Kleene closure modifies the automaton so that the set of symbol sequences or transductions is sequentially repeated zero or more times. Given an automaton A, the Kleene closure of A is denoted as A^*. The union combines two automata in parallel so that the resulting automaton represents a union of the two sets for the two automata. The concatenation combines two automata in series. The resulting automaton represents a set of symbol sequences or transductions, in which each element is a concatenation of two elements in the two respective automata.

Given two automata A_1 and A_2, their union is denoted as $A_1 \cup A_2$, and their concatenation is denoted as $A_1 \cdot A_2$. Examples of the original and resulting WFSTs for Kleene closure, union and concatenation are shown in Fig. 3.6. These operations can also be applied to FSAs, FSTs, and WFSAs.

With transducers, *projection*, *inversion*, and *composition* are important operations. Projection is an operation for converting a transducer into an acceptor by omitting input or output labels from the transducer. Inversion is an operation for inverting the input and output labels at each transition. The resulting transducer represents a set of inverted transductions. Figure 3.7 shows the results of the projection and inversion of the WFST T_A in Fig. 3.6 (a). Composition is an operation for combining two transducers into one single transducer that represents a set of transductions cascaded with the original two transducers. In addition, there are several optimization operations.

In the succeeding sections we provide further details about composition and some optimization operations. However, first we define terms for describing their concepts and algorithms.

Given a WFST $T = (\Sigma, \Delta, Q, I, F, E, \lambda, \rho)$, for any state $q \in Q$, we denote a multi-set of transitions outgoing from q by $E[q]$, and for any transition $e \in E$, we denote its input label by $i[e]$, its output label by $o[e]$, its origin state by $p[e]$, its destination state by $n[e]$, and its weight by $w[e]$. We denote a *path* as a series of consecutive transitions $\pi = e_1, \ldots, e_k$ such that $n[e_{j-1}] = p[e_j]$ for $j = 2, \ldots, k$. We extend $n[\cdot]$ and $p[\cdot]$ for paths as $n[\pi] = n[e_k]$ and $p[\pi] = p[e_1]$. We also extend $o[\cdot]$ and $w[\cdot]$ as $o[\pi] = o[e_1] \cdot \cdots \cdot o[e_k]$ and $w[\pi] = w[e_1] \otimes \cdots \otimes w[e_k]$. If $n[e_1] \in I$ and

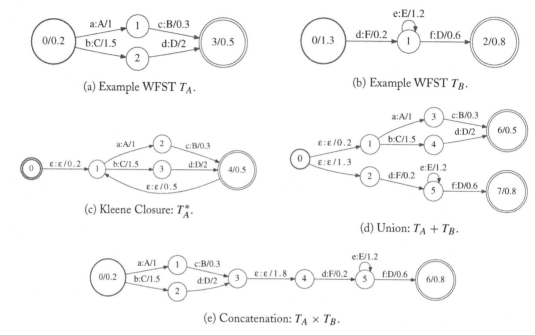

(a) Example WFST T_A.

(b) Example WFST T_B.

(c) Kleene Closure: T_A^*.

(d) Union: $T_A + T_B$.

(e) Concatenation: $T_A \times T_B$.

Figure 3.6: Examples of original and resulting WFSTs for rational operations

(a) Projection of T_A.

(b) Inversion of T_A.

Figure 3.7: Examples of projection and inversion

$n[e_k] \in F$, the path is said to be a *successful path*. If a state is accessible from both an initial and a final state, the state is denoted *coaccessible*. All the coaccessible states are on at least one successful path. States that are not coaccessible are termed *dead states*. Similarly, transitions that are not coaccessible are called *dead transitions*. Removing all the dead states and transitions from an FA is called *trimming*, and an FA without any dead states and transitions is called a *trimmed* FA, which is obtained as a result of trimming.

3.5   TRANSDUCER COMPOSITION

We explain the principle of composition and its algorithm. Transducers shown in Fig. 3.8 (a) and (b) assume that the transducer in (a) converts a sequence of letters so that they are all uppercase letters, and the transducer in (b) converts the sequence of uppercase letters to a sequence of specific words that matches the letters. The composition of these transducers results in the transducer shown in Fig. 3.8 (c), which converts a sequence of letters into the word sequence. Actually, it is not easy to construct the resulting transducer from scratch because we need to consider uppercase and lowercase letters for each word entry. However, complicated transductions can often be factored into a cascade of more simple transductions. In many cases, it is much easier to design simple component transducers and then combine them than to design one directly from scratch.

First we show Algorithm 10, which is an epsilon-free composition algorithm for WFSTs which combines two WFSTs such that the first transducer does not have any epsilon outputs and the second does not have any epsilon inputs. We will then show how to compose general transducers including epsilons.

In the algorithm, transducers $T_1 = (\Sigma_1, \Delta_1, Q_1, I_1, F_1, E_1, \lambda_1, \rho_1)$ and $T_2 = (\Sigma_2, \Delta_2, Q_2, I_2, F_2, E_2, \lambda_2, \rho_2)$ are combined into $T = (\Sigma_1, \Delta_2, Q, I, F, E, \lambda, \rho)$. Each state in T is composed by coupling a state in T_1 and a state in T_2. Therefore, a composite state is identified as a pair of original states, (q_1, q_2), where $q_1 \in Q_1$ and $q_2 \in Q_2$. Each transition from the composite state (q_1, q_2) is also made by coupling transitions e_1 and e_2 such that $e_1 \in E[q_1]$, $e_2 \in E[q_2]$, and $o[e_1] = i[e_2] \neq \varepsilon$. This results in a transition that has input $i[e_1]$, output $o[e_2]$, and weight $w[e_1] \otimes w[e_2]$. In this way, T becomes a transducer directly performing transductions that are cascaded by T_1 and T_2. The computational complexity of this algorithm is $O(|T_1||T_2|)$, where $|T_i| = |Q_i| + |E_i|$ for $i = 1$ or 2.

First, the algorithm composes initial states on lines 1-4, where a set of initial states is obtained as a direct product of I_1 and I_2. The initial weight for each initial state (i_1, i_2) is obtained as $\lambda(i_1) \otimes \lambda(i_2)$. The initial states are then inserted in a queue on line 5. Next it composes transitions from each composed state (q_1, q_2) in the queue. In lines 9-12, (q_1, q_2) becomes a final state in T if both q_1 and q_2 are final states in T_1 and T_2. The final weight $\rho((q_1, q_2))$ is computed as $\rho(q_1) \otimes \rho(q_2)$. In lines 13-19, a transition from state (q_1, q_2) is made for each transition pair $(e_1, e_2) \in E[q_1] \times E[q_2]$ such that $o[e_1] = i[e_2]$, and the next state $(n[e_1], n[e_2])$ as a pair of next states for e_1 and e_2 is made for the transition, and then inserted in the queue at line 16 for further composition. As a result, the following composite transition is obtained; $((q_1, q_2), i[e_1], o[e_2], w[e_1] \otimes w[e_2], (n[e_1], n[e_2]))$, which is added into E by a join operation $\uplus$ in line 19. Since E is a multi-set, transitions are allowed to appear more than once. In terms of composition in general, the operation needs a special treatment for epsilons, where the epsilon-free algorithm is extended by simulating the epsilon transitions. This extension can be explained with FST level operations. First we modify T_1 and T_2 to T_1' and T_2'. As mentioned in Section 3.2, an epsilon transition is made without any input. For composition, we need to consider two cases, namely when T_1 makes a transition with an epsilon output and when T_2 makes a transition with an epsilon

(a) Letter transducer: transitions for some letters are omitted because of space limitations.

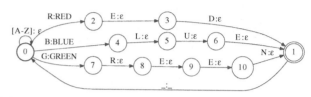

(b) Word recognition transducer that picks up words RED, BLUE, and GREEN.

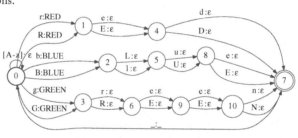

(c) Composition of (a) and (b).

Figure 3.8: Examples of component and composite transducers: for simplification we introduce a label [A-Z] (or [A-z]) that represents a set of English letters in the figure. The transition with this label actually represents multiple transitions for all the symbols in the set.

input. In the first case, T_1 outputs nothing and therefore only T_1 makes the epsilon transition and T_2 stays in the current state. In the second case, T_2 makes the epsilon transition with no input while T_1 stays in the current state. To simulate these state transitions, we expand T_1 and T_2 to T_1' and T_2' as in Fig. 3.9.

For T_1, the ε output is replaced with εo, and self loops that match εi are added to each state. For T_2, the ε input is replaced with εi, and self loops that accept εo are added to each state. With this extension, the epsilon-free composition can be used for $T_1' \circ T_2'$, where εi and εo are considered regular symbols. The result of this composition is shown in Fig. 3.10.

Algorithm 10 WFST-Composition(T_1, T_2)

1: **for** each $(i_1, i_2) \in I_1 \times I_2$ **do**
2:     $\lambda((i_1, i_2)) \leftarrow \lambda_1(i_1) \otimes \lambda_2(i_2)$
3:     $I \leftarrow I \cup \{(i_1, i_2)\}$
4: **end for**
5: $Q \leftarrow S \leftarrow I$
6: **while** $S \neq \emptyset$ **do**
7:     $(q_1, q_2) \leftarrow \text{Head}(S)$
8:     $\text{Dequeue}(S)$
9:     **if** $(q_1, q_2) \in F_1 \times F_2$ **then**
10:       $F \leftarrow F \cup \{(q_1, q_2)\}$
11:       $\rho((q_1, q_2)) \leftarrow \rho_1(q_1) \otimes \rho_2(q_2)$
12:    **end if**
13:    **for** each $(e_1, e_2) \in E[q_1] \times E[q_2]$ such that $o[e_1] = i[e_2]$ **do**
14:       **if** $(n[e_1], n[e_2]) \notin Q$ **then**
15:          $Q \leftarrow Q \cup \{(n[e_1], n[e_2])\}$
16:          $\text{Enqueue}(S, (n[e_1], n[e_2]))$
17:       **end if**
18:       $E \leftarrow E \uplus \{((q_1, q_2), i[e_1], o[e_2], w[e_1] \otimes w[e_2], (n[e_1], n[e_2]))\}$
19:    **end for**
20: **end while**
21: **return** $T = (\Sigma_1, \Delta_2, Q, I, F, E, \lambda, \rho)$

However, the transducer in Fig. 3.10 includes redundant paths. In addition, the weight summing over possible paths is actually incorrect for non-idempotent semirings such as probability and log semirings. To solve this problem, a special transducer called a *filter* is introduced in the composition operation. Assuming F is a filter, we perform $T_1' \circ F \circ T_2'$ instead of $T_1' \circ T_2'$. For example, a 2-state filter (*epsilon-sequencing filter*) in Fig. 3.11(a) can be used as F. In the figure, "x" represents any symbol in $\Sigma_2 \cap \Delta_1$.

The composition $T_1' \circ F \circ T_2'$ results in the transducer of Fig. 3.12(a). In the figure, dead states and transitions are shown in gray. Although such states and transitions are composed by the algorithm, they can later be removed by a trimming operation.[1] During composition with a 2-state filter, an epsilon output (εo) transition in T_1 is not allowed immediately after an epsilon input (εi) transition in T_2, because the filter changes state from 0 to 1 by εo of T_2, but does not accept εo of T_1 from state 1. Consequently, whenever both an epsilon output transition in T_1 and an epsilon input transition in T_2 are possible from each state paired in a composite state, the transition in T_1 is always located first in the composite transducer. Finally, the redundant paths are successfully filtered out.

---

[1]The trimming operation first detects dead states and transitions by checking the coaccessiblity from the initial and final states using a graph traversal technique, and then removes them from the original automaton.

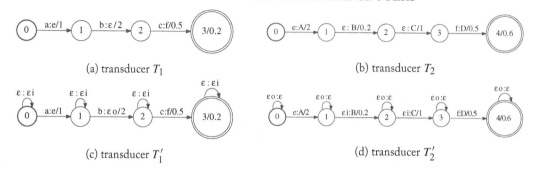

Figure 3.9: Transducers including epsilons and modified transducers for general composition

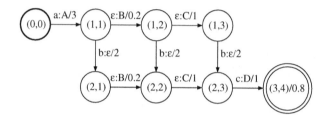

Figure 3.10: Result of composition $T_1' \circ T_2'$

On the other hand, we can use a 3-state filter (*epsilon-matching filter*) as in Fig. 3.11 (b), which derives the composite transducer in Fig. 3.12(b) for $T_1' \circ F \circ T_2'$. The 3-state filter makes it necessary to match the epsilon output of T_1' with the epsilon input of T_2' if possible, and therefore the resulting transducer has shorter paths than those obtained with the 2-state filter. A filter can be selected depending on the target composition. If we want to keep the same number of original epsilons on the composite paths, we should use a 2-state filter. If not, we can use a 3-state filter to obtain a more compact WFST.

In practical implementations of the composition operation, filter transducers are usually embedded directly in the program code, i.e., state transitions in the filter are simulated in the code without using the data structure for a general WFST. This improves the speed and memory efficiency of the composition operation.

3.6 OPTIMIZATION

Here we briefly introduce important optimization operations, determinization, weight pushing, and minimization for WFSTs. These operations transform the structure of a WFST into a more efficient one in terms of speed and memory in run time, but they retain the function represented by the original WFST.

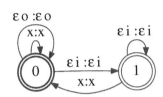

(a) 2-state filter (epsilon-sequencing filter).

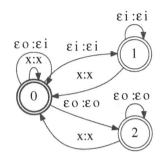

(b) 3-state filter (epsilon-matching filter).

Figure 3.11: Filter transducers

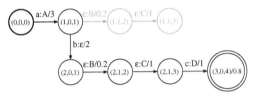

(a) Composite WFST with 2-state filter.

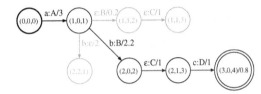

(b) Composite WFST with 3-state filter.

Figure 3.12: Result of $T_1' \circ F \circ T_2'$

3.6.1 DETERMINIZATION

Determinization is the most important optimization operation for finite automata (FAs). The availability of determinization is a big advantage as regards solving sequence recognition or transduction problems using FAs, because it yields the most efficient FA in terms of speed, i.e., a deterministic FA (DFA) that is equivalent to the original FA. When designing an FA to solve a problem, we try to design it transparently so that it is easy to understand. However, such an FA is not always efficient in terms of speed when we run it on a computer. Determinization is useful for accelerating its run-time speed.

As mentioned in 3.2, each transition made by the DFA is unique at any state for a given input symbol, and therefore the successful path is also unique for an input symbol sequence if the sequence is accepted by the DFA. The computational complexity involved in finding a successful path is linear for the length of the sequence. The computation needed to find a transition with an input label that matches each input symbol is proportional to $\log_2 |E(q)|$, where $|E(q)|$ denotes the number of outgoing transitions from state q.

The pseudo-code of determinization for WFSTs is shown in Algorithm 11, which basically follows the classical determinization algorithm for FSAs. In the classical method, if there are multiple transitions with the same input label outgoing from a state, they are merged into one transition with

---

Algorithm 11 WFST-Determinization(T)

1: $i' \leftarrow \{(i, \varepsilon, \lambda(i)) | i \in I\}$
2: $\lambda'(i') \leftarrow \bar{1}$
3: $Q' \leftarrow S \leftarrow \{i'\}$
4: **while** $S \neq \emptyset$ **do**
5:   $p' \leftarrow \text{Head}(S)$
6:   $\text{Dequeue}(S)$
7:   **for** each $x \in \{x | i[e] = x, e \in E[p], p \in Q[p']\}$ **do**
8:     $y' \leftarrow \bigwedge\{z \cdot y | (p, z, v) \in p', (p, x, y, w, q) \in E\}$
9:     $w' \leftarrow \bigoplus\{v \otimes w | (p, z, v) \in p', (p, x, y, w, q) \in E\}$
10:     $q' \leftarrow \{(q, y'^{-1} \cdot z \cdot y, \bigoplus\{w'^{-1} \otimes v \otimes w | (p, z, v) \in p', (p, x, y, w, q) \in E\})$
       $| (p, z, v) \in p', (p, x, y, w, q) \in E\}$
11:     $E' \leftarrow E' \cup \{(p', x, y', w', q')\}$
12:     **if** $q' \notin Q'$ **then**
13:       $Q' \leftarrow Q' \cup \{q'\}$
14:       **if** $Q[q'] \cap F \neq \emptyset$ **then**
15:         $F' \leftarrow F' \cup \{q'\}$
16:         $\rho'(q') \leftarrow \{(z, \bigoplus\{v \otimes \rho(q) | (q, z, v) \in q', q \in F\}) | (q, z, v) \in q', q \in F\}$
17:       **end if**
18:       $\text{Enqueue}(S, q')$
19:     **end if**
20:   **end for**
21: **end while**
22: **return** $T' = (\Sigma, \Delta, Q', \{i'\}, F', E', \lambda', \rho')$

---

the same input label in the determinized FSA. The destination states of the multiple transitions are also merged into one state. Therefore, each state in the determinized FSA is identified by a subset of states in the original FSA. When making a transition with an input label from an already determinized state, all transitions with the input label outgoing from all the states in the subset associated with the determinized state are merged into a new transition, and their destination states are also merged. In this way, states and transitions of the determinized FSA can be constructed iteratively from its initial state, which is identified by the set of initial states in the original FSA.

With WFSTs, the algorithm needs to be extended to deal with output labels and weights.[2] Roughly speaking, the classical determinization binds the transitions and destination states by input label. However, WFSTs have an output label and a weight on each transition. Even if the input labels are the same in the transitions, their output labels and weights may differ. Basically, different

---

[2]A general determinization algorithm for WFSAs is presented in [Moh09]. The algorithm for WFSTs can be explained as a special case in which a pair of an output string and a weight is considered a single weight. However, we here focus on an algorithm dedicated to WFSTs for consistency in terms of the algorithms presented in this book.

output labels and weights cannot be assigned to one transition. Hence, in WFST determinization, a common prefix for the output labels and the sum of the weights over the transitions are assigned to the new transition. The leftover output label excluding the common prefix and the residual weight excluding the sum of the weights for each original transition are held in the new destination state together with the original destination states. Accordingly, each determinized state is identified by a set of triplets, $\{(p, z, v)|p \in Q, z \in \Delta^*, v \in \mathbb{K}\}$, where p is a state in the original WFST, z is a leftover output label, and v is a residual weight. To perform this operation, the semiring needs to be weakly left-divisible for dividing the sum of the weights from each transition weight.

Algorithm 11 generates a deterministic WFST $T' = \langle \Sigma, \Delta, Q', I', F', E', \lambda', \rho' \rangle$ from the original WFST $T = \langle \Sigma, \Delta, Q, I, F, E, \lambda, \rho \rangle$. It starts by generating one initial state i' for T' on line 1, where i' corresponds to a set of triplets, $\{(i, \varepsilon, \lambda(i))|i \in I\}$. Then, i' is inserted in a queue S, and also inserted in Q' on line 3. In lines 7-20, new states are derived from state p' that is taken out of S. For each input label x in the set of input labels for transitions leaving $Q[p']$, all the original states belonging to p', a new state q' and a transition (p', x, y', w', q') are made in lines 8-11. Over triplets of (p, z, v) in p' and transitions of (p, x, y, w, q) in $E[p]$, a set of strings, $\{z \cdot y\}$ is made in line 8. Then, y' is obtained as the common head symbol in the set, which is denoted by $\bigwedge$.[3] Here we assume that the common head symbol equals ε if there is no common head symbol in the set. The weight w' can also be computed as the $\oplus$-sum of $v \otimes w$ over the triplets and the transitions at line 9. On line 10, a set of triplets that includes the leftover string and the residual weight is defined as the new state q'. If q' is not included in Q', it is added to Q' in lines 12-13.

Lines 14-17 deal with final states, final output labels, and final weights of T'. If there is at least one final state in q', q' can be a final state. In line 16, $\rho'(q')$ is defined as a set of pairs consisting of a final output string and a final weight for state q', which correspond to the pairs of leftover strings and residual weights in q'. $\rho'(q')$ obviously has a different meaning from the final weight function found in the definition of WFST in Section 3.1. It is similar to a final emission function of output symbols, which is defined for p-subsequential transducers as in Fig. 3.3(d). To convert the resulting p-subsequential transducer into a WFST according to the form shown in Fig. 3.3(e), the algorithm requires more steps, but we skip them to avoid complication. On line 18, the new state q' is pushed into the queue S. The above steps are repeated until S is not empty, and then T' is returned as a result of determinization of T on line 19.

Figure 3.13 shows an example of determinization. Given a non-deterministic WFST T as in Fig. 3.13(a), states realized by WFST determinization are generated as in Fig. 3.13(b). In this example, first the initial state is formed as $\{(0, \varepsilon, 0.2)\}$ whose weight is 0.2. Then, a transition from $\{(0, \varepsilon, 0.2\}$ and its destination state are generated based on the two original transitions, which originate from state 0, and their input symbol is "a," i.e., $(0,a,X,0.5,1)$ and $(0,a,Y,1.2,2)$ in the original WFST. The output label of the new transition is obtained as $X \wedge Y = \varepsilon$. The transition weight is $0.5 \oplus 1.2 = 0.5$ in a tropical semiring. The destination state becomes $\{(1, X, 0), (2, Y, 0.7)\}$ by

---

[3] According to the algorithm of [Moh09], the output label for the transition, y', is obtained as the longest common prefix in the set of such strings. However, since y' should be a single symbol in our case, we just pick at most one common head symbol of them.

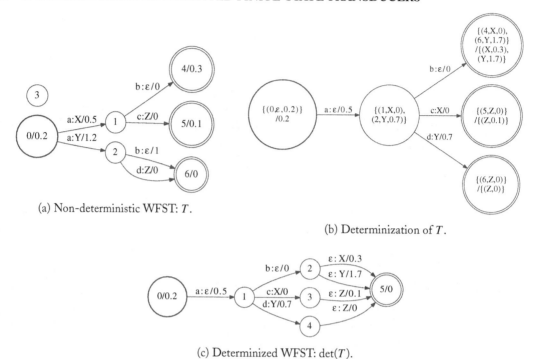

(a) Non-deterministic WFST: T.

(b) Determinization of T.

(c) Determinized WFST: $\det(T)$.

Figure 3.13: Example of WFST determinization

grouping states 1 and 2 of the original WFST. Subsequently, states 4 and 6 are grouped based on the transitions with input symbol "b," and become $\{(4, X, 0), (6, Y, 1.7)\}$ with the final emission $\{(X, 0.3), (Y, 1.7)\}$. Other states and transitions in the figure are also constructed with the same procedure. After this determinization process, all the states are renumbered and the final emission function in each final state is replaced with epsilon transitions to a new final state as the WFST of $\det(T)$ in Fig. 3.13(c).

3.6.2   WEIGHT PUSHING

The weight pushing operation moves the weights distributed over all the paths to the initial states in a WFSA or a WFST without changing its function. In many sequence recognition or transduction problems, finding the most likely or minimal cost solution is a main task for solving the problem. When we apply a weighted automaton, the problem is interpreted as being a search for the minimally or maximally weighted successful path in the automaton. It is well known that the effect of weight look-ahead caused by pushing weights accelerates the search process since the unpromising paths can be eliminated in the early stage of the search, and therefore the total processing time can be reduced.

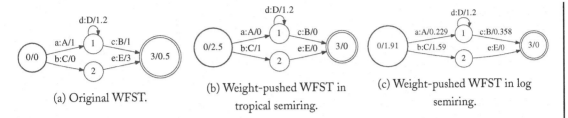

(a) Original WFST.   (b) Weight-pushed WFST in tropical semiring.   (c) Weight-pushed WFST in log semiring.

Figure 3.14: Weight pushing

Figure 3.14 shows a WFST and the results of weight pushing in a tropical semiring and a log semiring. In Fig. 3.14(a), the weight of the successful path along states 0,1,3 is obtained as $1 \otimes 1 \otimes 0.5 = 2.5$. All the path weight is moved to the initial state in Fig. 3.14(b). The other weight for the path of states 0,2,3 is $0 \otimes 3 \otimes 0.5 = 3.5$, and this is pushed to the initial state with weight 2.5 and to the first transition from 0 to 2 with weight 1 in Fig. 3.14(b). Thus, the weight of each successful path in the original WFST is retained in the weight-pushed WFST.

A general weight pushing algorithm comprises two steps. The first step computes a *potential* for each state, which is obtained as the sum of the weights over all the paths originating from that state to some final states. In the second step, the weight for each transition is modified with the potential difference between the source and destination states of the transition. As a result, the weights move to the initial states. Note that each transition weight is decided based on the difference between potentials, i.e., the bias component is removed and moved to the initial states. A set of potentials for all the states can be obtained with a general shortest distance algorithm from the final states, which is given as Algorithm 12.

In the algorithm, $V[q]$, the potential for state q, is computed as

$$V[q] = \bigoplus_{\pi \in \Pi(q,F)} w[\pi] \otimes \rho(n[\pi]), \tag{3.1}$$

where $\Pi(q, F)$ is a set of paths that originate from state q and arrive at a final state in F. $w[\pi]$ is the path weight of path π. However, we need to consider that $\Pi(q, F)$ is infinite if cycles are included in the paths from q to a final state. Since it is impossible to deal with infinite paths, the algorithm postulates that the semiring is k-closed. Accordingly, for any cycle c, the equation

$$\bigoplus_{n=0}^{l} w[c]^n = \bigoplus_{n=0}^{k} w[c]^n$$

is satisfied for any integer l such that $l > k$. Thus, we can omit l or more loops in each cycle from the sum of those path weights.

In lines 1-7 of Algorithm 12, potentials $V[q]$ are initialized by $\rho(q)$ for the final states and $\bar{0}$ for the other states. In line 8, the final states in F are first enqueued in S. In lines 9-23, the potentials are computed based on a backward search from the final states. In lines 10-13, state q is removed

---

Algorithm 12 WFST-Potential(T)

1: **for** each $q \in Q$ **do**
2:    **if** $q \in F$ **then**
3:       $V[q] \leftarrow r[q] \leftarrow \rho(q)$
4:    **else**
5:       $V[q] \leftarrow r[q] \leftarrow \bar{0}$
6:    **end if**
7: **end for**
8: $S \leftarrow F$
9: **while** $S \neq \emptyset$ **do**
10:     $q \leftarrow \text{Head}(S)$
11:     $\text{Dequeue}(S)$
12:     $R \leftarrow r[q]$
13:     $r[q] \leftarrow \bar{0}$
14:     **for** each $e \in E^{-1}[q]$ **do**
15:        **if** $V[p[e]] \neq V[p[e]] \oplus (R \otimes w[e])$ **then**
16:           $V[p[e]] \leftarrow V[p[e]] \oplus (R \otimes w[e])$
17:           $r[p[e]] \leftarrow r[p[e]] \oplus (R \otimes w[e])$
18:           **if** $p[e] \notin S$ **then**
19:              $\text{Enqueue}(S, p[e])$
20:           **end if**
21:        **end if**
22:     **end for**
23: **end while**
24: **return** V

---

from S, $r[q]$ is recorded in R and $r[q]$ is set at $\bar{0}$, where $r[q]$ means the sum of the path weights added since the last time q was popped. In lines 14-22, for each transition e incoming to state q, the path weight is propagated to its previous state $p[e]$, where $E^{-1}[q]$ represents a multi-set of incoming transitions to q. In line 15, the current potential $V[p[e]]$ for state $p[e]$ is compared with the new potential, i.e., $V[p[e]] \oplus (R \otimes w[e])$. If the current and the new potentials are equal, $V[p[e]]$ does not have to be updated and not enqueued. If they are not equal, the potential is updated and state $p[e]$ is enqueued for the further summing up of path weights. When S becomes empty, i.e., all the potentials are fixed, the algorithm terminates and returns the set of potentials in line 24.

This algorithm is guaranteed to terminate for k-closed semirings. The tropical semiring is 0-closed. The log semiring is k-closed if we disregard small differences between weights. For such k-closed semirings, the condition on line 15 needs to admit small differences between them, i.e., we assume $x = y$ if $|x - y| < \delta$ where δ is a small value that can be ignored. In addition, the algorithm

---

Algorithm 13 WFST-WeightPushing(T)

1: $V[\cdot] \leftarrow$ WFST-Potential(T)
2: **for** each $q \in Q$ **do**
3:    **if** $q \in I$ **then**
4:       $\lambda(q) \leftarrow \lambda(q) \otimes V[q]$
5:    **end if**
6:    **for** each $e \in E[q]$ **do**
7:       $w[e] \leftarrow V[q]^{-1} \otimes w[e] \otimes V[n[e]]$
8:    **end for**
9:    **if** $q \in F$ **then**
10:       $\rho(q) \leftarrow V[q]^{-1} \otimes \rho(q)$
11:    **end if**
12: **end for**

---

admits any queue discipline for S. A priority queue in which the state with the largest potential is popped first is often chosen, because potentials become larger and fixed earlier.

The weight pushing operation updates the weights using the potentials. As in Algorithm 13, the initial weight $\lambda(q)$, transition weight $w[e]$, and final weight $\rho(q)$ are updated as in lines 4, 7, and 10, respectively. In the weight pushing algorithm, it is assumed that WFST T is a trimmed FA and the weight semiring is weakly left-divisible and zero-sum free such as tropical and log semirings.

3.6.3   MINIMIZATION

Minimization is an algorithm designed to minimize the number of states for any DFA. The minimization for WFSTs is described in [Moh09], and consists of two steps:

1. Push weights and output labels to the initial states in the WFST, and

2. Minimize the WFST using a classical minimization algorithm assuming that the triplet "input:output/weight" on each transition is one single label.

There are several algorithms for minimization [BBCF10]. Such algorithms basically obtain a partition of the set of states in the automaton, i.e., the set of states is divided into non-overlapping and non-empty blocks. The partition is determined so that each block includes equivalent states that are not distinguished from each other, i.e., they accept the identical set of symbol sequences along the paths from those states to some final states. With WFSTs, two states $q_1, q_2 \in Q$ are equivalent if and only if

$$L(q_1, x) = L(q_2, x), \forall x \in \Sigma^*, \tag{3.2}$$

where

$$L(q, x) = (o[\pi], w[\pi]) \; s.t. \; p[\pi] = q, i[\pi] = x, n[\pi] \in F. \tag{3.3}$$

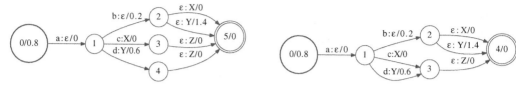

(a) Weight-pushed WFST: push(det(T)), where det(T) is of Fig. 3.13 (c).

(b) Minimized WFST: min(push(det(T))).

Figure 3.15: Example of minimization

We assume here that the WFST is deterministic, i.e., given q and x, path π is unique.

Once the partition is obtained, all the states are replaced with a new set of states each of which corresponds to a block of equivalent states in the partition. Transitions outgoing from all states in a block are bound by their labels, and they are redirected for the new states, where the state indices are replaced with those for the blocks to which the original states belong.

Figure 3.15(a) shows the result of push(det(T)), where weight pushing is applied to det(T) in Fig. 3.13(c). Figure 3.15(b) shows the result of min(push(det(T))), i.e., the minimization of push(det(T)). Finally, states 3 and 4 of det(T) are considered to be equivalent, and merged into state 3 in min(push(det(T))). Note that states 3 and 4 are not equivalent in det(T), and therefore weight pushing is needed to minimize the WFST. Label pushing is also applied as necessary. For example, it can be realized by considering each output label as a weight in a string semiring.

Hopcroft's algorithm is a well-known algorithm for efficiently minimizing DFAs, whose computational complexity is $O(|E| \log |Q|)$. This algorithm iterates the division of blocks, starting from a smaller number of larger blocks, until every block becomes an equivalent state set. To split each block in the current partition, one block is chosen as a *splitter* with which each block is split and the partition is updated. At each splitting step, a subset of states in the block, which have transitions with a label to some states in the splitter, is taken out of the block. Then, the original block is replaced with the new blocks, i.e., the subset and the remainder. By iterating this process, the partition is gradually refined into a set of equivalent state subsets. The proof can be seen in many studies.

Algorithm 14 shows the minimization algorithm slightly modified for WFSTs. On the first line, partition $\mathcal{P}$ and queue $\mathcal{W}$ are initialized with $\emptyset$. The queue $\mathcal{W}$ is a waiting list used for containing the blocks used to split each block in $\mathcal{P}$. In lines 2-7, an initial partition is derived so that $\mathcal{P}$ includes blocks of final states with the same final weight, and one for non-final states. In line 5, blocks of final states are inserted into $\mathcal{W}$ as splitters.

The original Hopcroft's algorithm initializes partition $\mathcal{P}$ as two blocks F and $Q - F$. For the weighted transducers, we further split F according to their final weights because final states with different weights are obviously not equivalent. In addition, we may obtain a finer initial partition by using information of outgoing transitions from each state because two states are differentiated from each other if the two states have different sets of outgoing transitions as regards triplet of input,

---

Algorithm 14 WFST-Minimization(T)

1: $\mathcal{P} \leftarrow \mathcal{W} \leftarrow \emptyset$
2: **for** each $\rho \in \{\rho(f) | f \in F\}$ **do**
3:      $F_\rho \leftarrow \{f | \rho(f) = \rho, f \in F\}$
4:      $\mathcal{P} \leftarrow \mathcal{P} \cup \{F_\rho\}$
5:      Enqueue($\mathcal{W}, F_\rho$)
6: **end for**
7: $\mathcal{P} \leftarrow \mathcal{P} \cup \{Q - F\}$
8: **while** $\mathcal{W} \neq \emptyset$ **do**
9:      $S \leftarrow$ Head($\mathcal{W}$) ; Dequeue($\mathcal{W}$)
10:     **for** each $(i, o, w) \in \{(i[e], o[e], w[e]) \mid e \in E^{-1}[S]\}$ **do**
11:         $R_{i,o,w} \leftarrow \{p[e] \mid i[e] = i, \ o[e] = o, \ w[e] = w, \ e \in E^{-1}[S]\}$
12:         **for** each $B \in \mathcal{P}$ such that $B \cap R_{i,o,w} \neq \emptyset$ and $B \not\subseteq R_{i,o,w}$ **do**
13:             $B_1 \leftarrow B \cap R_{i,o,w}$
14:             $B_2 \leftarrow B - B_1$
15:             $\mathcal{P} \leftarrow (\mathcal{P} - \{B\}) \cup \{B_1, B_2\}$
16:             **if** $B \in \mathcal{W}$ **then**
17:                 Erase($\mathcal{W}, B$); Enqueue($\mathcal{W}, B_1$); Enqueue($\mathcal{W}, B_2$)
18:             **else**
19:                 **if** $|B_1| \leq |B_2|$ **then**
20:                     Enqueue($\mathcal{W}, B_1$)
21:                 **else**
22:                     Enqueue($\mathcal{W}, B_2$)
23:                 **end if**
24:             **end if**
25:         **end for**
26:     **end for**
27: **end while**
28: $Q' \leftarrow \mathcal{P}$
29: **for** each $e \in E$ **do**
30:     $E' \leftarrow E' \cup \{(B(p[e]), i[e], o[e], w[e], B(n[e]))\}$
31: **end for**
32: **for** each $S \in Q'$ such that $S \subseteq F$ **do**
33:     $F' \leftarrow F' \cup \{S\}$
34:     $\rho'(S) \leftarrow \rho(q)$ for some $q \in S$
35: **end for**
36: **return** $T' = (\Sigma, \Delta, Q', I, F', E', \lambda, \rho')$

---

output, and weight on each transition. Hence, we can further classify all the states into smaller blocks and construct a finer partition. This initial partition is possible to accelerate the main steps of the algorithm. But we do not describe this initialization because of space limitations.

The procedure in lines 8-27 iteratively splits the blocks. In line 9, block S is removed from queue $\mathcal{W}$. In line 10, a set of triplets of input, output, and weight for transitions in $E^{-1}[S]$ is prepared, where $E^{-1}[S]$ denotes a set of transitions incoming to at least one state in S, i.e., $\{e \in E | n[e] \in S\}$. For each triplet (i, o, w) in the set, a binary split for each block is performed in lines 11-25. $R_{i,o,w}$ in line 11 is obtained as a set of previous states of transitions into some state in block S with (i, o, w). Each block B in $\mathcal{P}$ can be split by $R_{i,o,w}$ into $B_1 = B \cap R_{i,o,w}$ and $B_2 = B - B_1$ in lines 12 and 13. B is removed from $\mathcal{P}$, and new blocks B_1 and B_2 are added to $\mathcal{P}$ on line 15. The smaller one of the tow blocks B_1 and B_2 is enqueued in $\mathcal{W}$ to be used as a splitter.

After the right partition is obtained, the WFST is reconstructed for minimization in lines 28-35. In line 28, a minimal set of states, Q', is obtained as the partition $\mathcal{P}$. In lines 29-31, for each transition $e \in E$, the previous and next states are changed into the indices of blocks, where $B(s)$ returns the index of the block to which the state s belongs. In lines 32-35, final states and weights are decided for the new states.

Since Hopcroft's algorithm is general, it can be applied to any determinized WFSTs including cyclic ones. If the WFST is acyclic, we can use a more efficient algorithm called Revuz's algorithm [Rev92], which complexity is $O(|E|)$. In this method, the initial partition is obtained according to the maximum height from a final state using a depth-first search. Each block based on the maximum height is split according to the input label. See details in [Rev92].

3.7 EPSILON REMOVAL

Epsilon removal is an operation for converting an ε-NFA to its equivalent NFA without epsilon transitions. As mentioned in Section 3.2, the existence of epsilon transitions makes the FA non-deterministic. Although the determinization algorithm can be applied to ε-NFA by considering ε as a regular symbol, the resulting FA still has epsilon transitions and therefore the FA is not completely determinized. By performing the epsilon removal operation before determinization, the FA can be fully determinized.

Epsilon removal basically deletes epsilon transitions and adds new non-epsilon transitions, which are then spread from each state to all the states that can be reached with one or more epsilon transitions plus one non-epsilon transition. The input label of the new transition is set so that it is the same as that of the last non-epsilon transition. If there is a non-final state that can reach a final state solely with epsilon transitions, the state becomes a final state in the epsilon-removed automaton. Figure 3.16 shows the basic concept of epsilon removal. Figure 3.16 (a) is assumed to be a part of ε-NFA. The epsilon removal first finds an epsilon closure for each state. An epsilon closure is a set of states that can be reached from a given state by only epsilon transitions. The meshed states 1 and 2 in Fig. 3.16 (b) correspond to the elements in the epsilon closure for state 0. Then, to each state directed with a non-epsilon label from those in the epsilon closure, a transition with a non-epsilon

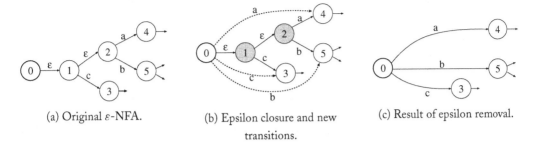

(a) Original ε-NFA.   (b) Epsilon closure and new transitions.   (c) Result of epsilon removal.

Figure 3.16: Concept of epsilon removal

label is added. In Fig. 3.16 (b), dashed lines correspond to those new transitions. Finally, the epsilon transitions are deleted, and all dead states and transitions are also removed. In Fig. 3.16 (c), the paths that were passing through epsilon transitions have been removed, and only non-epsilon transitions remain.

Output labels and weights need to be considered for WFSTs. An efficient method is presented in [Moh02, Moh09]. First, a pair of input and output epsilons, i.e., $\varepsilon : \varepsilon$, is assumed to be a single epsilon in the algorithm. Secondly, the distance from a source state to each destination state in the epsilon closure for the source state is computed, and it is multiplied by the weight of the last non-epsilon transition to obtain the weight of the new transition. The distance between the source and destination states can be obtained by a general single-source shortest distance algorithm along epsilon transitions.

Algorithm 15 shows the pseudo-code of epsilon removal for WFSTs. In lines 1-16, for each state p, epsilon transitions with $\varepsilon : \varepsilon$ are discarded and non-epsilon transitions are added so that the epsilon-removed WFST is equivalent to the original. First, only non-epsilon transitions are copied to E' (line 2). For each element (q, w') in the epsilon closure for state p, new transitions originating from state p are generated based on the non-epsilon transitions from state q, where the new transitions are given by $\{(p, x, y, w' \otimes w, r) | (q, x, y, w, r) \in E[q], (x, y) \neq (\varepsilon, \varepsilon)\}$ (line 3). In lines 5-14, the algorithm checks whether state p is final in the epsilon-removed WFST. State p is final if state p or q is final in the original WFST. If state q is final, the original final weight multiplied by w' is added to the final weight (line 13) because the path weight to state q with only epsilon transitions must be included. Finally, the epsilon-removed WFST T' is returned (line 17).

However, we need to consider that epsilon removal for transducers removes only transitions labeled with $\varepsilon : \varepsilon$. Therefore, transitions with $\varepsilon : x$ and $x : \varepsilon$ may remain in the transducer. An effective way of removing the maximum number of epsilons is to first apply *synchronization*. The synchronization converts a transducer so that it consumes input and output labels as synchronously as possible along each successful path. Therefore, transitions such as $\varepsilon : x$ and $x : \varepsilon$ decrease while those such as $(x : x)$ and $(\varepsilon : \varepsilon)$ increase. The epsilon removal excludes more epsilons from the transducer. The synchronization is explained in detail in [Moh09].

Algorithm 15 WFST-EpsilonRemoval(T)

1: **for** each $p \in Q$ **do**
2:      $E' \leftarrow E' \uplus \{e \in E[p] | (i[e], o[e]) \neq (\varepsilon, \varepsilon)\}$
3:      **for** each $(q, w') \in \text{EpsilonClosure}(p)$ **do**
4:          $E' \leftarrow E' \uplus \{(p, x, y, w' \otimes w, r) | (q, x, y, w, r) \in E[q], (x, y) \neq (\varepsilon, \varepsilon)\}$
5:          **if** $p \in F$ **then**
6:              $F' \leftarrow F' \cup \{p\}$
7:              $\rho'(p) \leftarrow \rho(p)$
8:          **end if**
9:          **if** $q \in F$ **then**
10:             **if** $p \notin F$ **then**
11:                 $F' \leftarrow F' \cup \{p\}$
12:             **end if**
13:             $\rho'(p) \leftarrow \rho'(p) \oplus (w' \otimes \rho(q))$
14:         **end if**
15:     **end for**
16: **end for**
17: **return** $T' = (\Sigma, \Delta, Q, I, F', E', \lambda, \rho')$

CHAPTER 4

Speech Recognition by Weighted Finite-State Transducers

In this chapter, we provide more details about the WFST approach to speech recognition. First we overview the approach, and then we explain how speech recognition models can be represented in WFST form and organized into a single search network. Finally, we show a time-synchronous Viterbi beam search algorithm when using a fully composed WFST.

4.1 OVERVIEW OF WFST-BASED SPEECH RECOGNITION

The WFST offers a unified form that can represent various knowledge sources utilized in state-of-the-art Large-Vocabulary Continuous-Speech Recognition (LVCSR), e.g., Hidden Markov Models (HMMs), phonotactic networks, lexical descriptions, and N-gram language models. And, multiple WFSTs, each of which corresponds to one such knowledge source, can be integrated into a fully composed single WFST that organizes an entire search network represented at the HMM-state level. Then the single WFST is converted into an equivalent and more efficient WFST by optimization methods, which eliminate the redundancy of the search network and therefore accelerate the decoding process. Here we overview WFST-based speech recognition.

As stated in Chapter 2, continuous speech recognition is defined as the problem of finding the most likely word sequence $\hat{W}$ for a given input speech O as in Eq. (2.3). The likelihood is calculated as $p(O|W)P(W)$ for any word sequence hypothesis W, where $p(O|W)$ is the acoustic likelihood and $P(W)$ is the language probability of W. More practically, a pronunciation probability $P(V|W)$ is incorporated, which is the probability of phone sequence V given W. These likelihoods and probabilities are calculated with given acoustic, pronunciation, and language models. Then Eq. (2.3) is rewritten as

$$\hat{W} = \operatorname*{argmax}_{W \in \mathcal{W}} \sum_{V \in R(W)} p(O|V, W)P(V|W)P(W) \qquad (4.1)$$

$$\approx \operatorname*{argmax}_{W \in \mathcal{W}} \left\{ \sum_{V \in R(W)} p(O|V)P(V|W)P(W) \right\}, \qquad (4.2)$$

where $p(O|V)$, $P(V|W)$, and $P(W)$ are calculated with the acoustic model, the pronunciation lexicon, and the language model, respectively. $\mathcal{W}$ is a set of possible word sequences and $R(W)$ is a set of possible phone sequences for word W. Since subword-based acoustic models are employed in state-of-the-art LVCSR systems, the acoustic likelihood $p(O|V, W)$ is assumed to depend only on the phone sequence V and therefore is approximated to $p(O|V)$.

Mainly for implementation reasons, we use the logarithms of these values in a Viterbi decoder. Namely the decoder performs the following with given models:

$$
\hat{W} \approx \operatorname*{argmax}_{W \in \mathcal{W}} \left\{ \max_{V \in R(W)} p(O|V) P(V|W) P(W) \right\} \tag{4.3}
$$

$$
= \operatorname*{argmax}_{W \in \mathcal{W}} \left\{ \max_{V \in R(W)} \{\log p(O|V) + \log P(V|W) + \log P(W)\} \right\}, \tag{4.4}
$$

where the summation is replaced with the maximum value obtained by the Viterbi approximation. For simplicity, henceforth we refer to the log likelihood and log probabilities as scores, i.e., $\log p(O|V)$ is referred to as the acoustic score, $\log P(V|W)$ as the pronunciation score and $\log P(W)$ as the language score. The WFST framework gives us an efficient procedure with which to solve Eq. (4.4).

In the WFST framework, the speech recognition problem is treated as a transduction from input speech signal O to a word sequence W. Each of the models used in speech recognition is interpretable as a WFST whose weights are defined as the negatives of scores. Namely, we consider WFSTs H, L and G corresponding to Eq. (4.4), which transduce O into V with weight $w_H(O \rightarrow V) = -\log p(O|V)$, V into W with weight $w_L(V \rightarrow W) = -\log P(V|W)$, and W into W with weight $w_G(W \rightarrow W) = -\log P(W)$, respectively. Then, the target transduction from O to W can be achieved by a cascade consisting of H, L and G. To make the transduction more efficient, these WFSTs are combined into one single WFST N that transduces O into W directly:

$$
N = H \circ L \circ G, \tag{4.5}
$$

where "$\circ$" is a composition operator. Then, the speech recognition is formulated as a search process on N for the word sequence with the minimum overall weight:

$$
\hat{W} \approx \operatorname*{argmin}_{W \in \mathcal{W}} \left\{ \min_{V \in R(W)} \{(-\log P(O|V)) + (-\log P(V|W)) + (-\log P(W))\} \right\} \tag{4.6}
$$

$$
= \operatorname*{argmin}_{W \in \mathcal{W}} \left\{ \min_{V \in R(W)} \{(w_H(O \rightarrow V) \otimes w_L(V \rightarrow W) \otimes w_G(W \rightarrow W)\} \right\}
$$

$$
= \operatorname*{argmin}_{W \in \mathcal{W}} w_N(O \rightarrow W). \tag{4.7}
$$

Note that this is obviously equivalent to the word sequence with the maximum overall scores in Eq. (4.4). Here we assume the weights are dealt with in a tropical semiring, i.e., "$\otimes$" performs the numerical addition "$+$."

When we incorporate triphone models, which are standard models in state-of-the-art speech recognition, we insert an additional WFST C that transduces a triphone sequence into a phone sequence:

$$N = H \circ C \circ L \circ G. \tag{4.8}$$

This fully composed WFST organizes an all-in-one search network, where the cross-word triphone contexts are exactly incorporated, which is complicated when we employ traditional approaches. The fully composed WFST can be optimized further by operations commonly employed in WFST frameworks, such as weighted determinization and minimization. These operations achieve optimization over the entire search space, and can greatly increase the search efficiency while the effects of similar techniques in traditional approaches have been limited to a local part of the search space basically corresponding to each model.

Once a fully composed WFST is constructed, the decoder works to find the best path in the WFST for any speech input. The WFST does not need to be updated unless the underlying models are updated. Thus, the decoder can concentrate on its search process using the optimized static search network unlike some conventional decoders that require the dynamic expansion of a less optimized search network. This is a primary advantage of the WFST framework over traditional approaches. In addition, since the decoder is designed to work with any WFST, the decoder program can be largely independent of speech models contained in the WFST. This is another advantage of the WFST framework, which allows the decoder to be used for more general purposes and easily maintained.

In the following sections, we describe how to construct component WFSTs for speech models, the composition and optimization steps needed to organize a fully composed WFST, and a decoding algorithm using the WFST for speech recognition. As mentioned above, we treat WFST-based speech recognition as a transduction from input speech signal to a word sequence. However, we should note that this transduction actually falls outside the definition of a WFST. As with traditional speech recognizers, the speech signal is converted into a sequence of real-valued vectors by feature extraction. But these vectors cannot be defined as elements of a finite set of symbols for WFSTs. In addition, the acoustic likelihood for such a vector cannot be embedded as a transition weight in advance because the likelihood must be calculated on demand using probability density functions (PDFs) of HMM states during decoding. Accordingly, the WFST-based speech recognizer must be separated into a part for handling speech input and another part for handling a pure WFST. We also make this point clear in the next section.

4.2   CONSTRUCTION OF COMPONENT WFSTS

In this section, we show how the knowledge sources for speech recognition are represented in WFST form, which are acoustic models, phone context dependency, pronunciation lexicons, and language models. Specifically, we focus on standard models used in speech recognition such as hidden Markov models, triphone context dependency, a simple (non probabilistic) pronunciation lexicon, and an n-gram language model.

4.2.1   ACOUSTIC MODELS

In the context of a WFST approach, a set of acoustic models can be viewed as a transducer that converts an input speech signal into a sequence of (context-dependent) phone units with the weight of the acoustic likelihood. Suppose a set of context-dependent HMMs for speech recognition is represented as a transducer as in Fig. 4.1. In the figure, three models for context-dependent phones s(t), (s)s, and (s)s(t) are included and connected so that they can be repeated via state 0 as the origin, where the label "(s)s(t)" denotes phone /s/ in the left context of /s/ and the right context of t. If there is no bracket on the left or right side, this means that the side is context independent. Each model is a left-to-right HMM with three states, and each state has one self loop and one outgoing transition. x is a meta symbol that represents the set of all vectors in the feature space of speech signals. The label "x : s(t)/w(x|S0)" for the state transition from 0 to 1 means that any input feature vector can be accepted and converted into an output symbol "s(t)" with weight function w(x|S0), where "S0" indicates the 0-th shared state in the HMMs. All the shared states have individual output probability density functions like $b_{Sk}(x)$ where Sk indicates the k-th shared state. In a tropical or log semiring, the weight function w(x|Sk) equals $-\log b_{Sk}(x)$. On the other hand, another transition holds an ε output label and a transition weight, which includes a minus log value of the state transition probability for the HMM. For example, the self loop at state 1 has weight $0.22 \otimes w(x|S0)$, where 0.22 corresponds to the minus log of the state transition probability of the HMM, and where the original probability is assumed to be 0.8. "$\otimes$" means the multiplication in the tropical or log semiring. Accordingly, the transducer accepts a sequence of feature vectors and transduces it into the context dependent phone sequence with the acoustic weight. Note that the context dependency of these models is not considered in this example, and is given by the context dependency transducer C.

Here, we should note that the transducer in Fig. 4.1 is outside the definition of a WFST, and therefore cannot be dealt with directly in the WFST framework. As mentioned in Section 4.1, the input feature vector consists of continuous-valued elements that are not supported in the definition of WFSTs. Hence, we needed to introduce meta symbol x and weight function w(x|Sk) to represent the HMMs as a transducer. To deal with the acoustic models in WFST-based speech recognition, the transducer is factorized into two parts, the HMM topology and acoustic matching. The former can be dealt with as a pure WFST, and the latter is dealt with virtually as a transducer in the decoder program. Figure 4.2 shows an example of the factorization for the transducer in Fig. 4.1. The HMM topology WFST (1) accepts a sequence of labels each of which corresponds to a shared state in the HMMs. Thus, the HMM topology can be written as a formal WFST. On the other hand, the acoustic matching transducer (1) holds x and w(x|Sk) and these computations are performed on demand and combined with the other WFST by using a dedicated program in the decoder.

Another type of factorization is also possible. Figure 4.3 and 4.4 are such examples. In Fig. 4.3, the HMM topology simply represents the order of the states. Self loops and state transition probabilities are included in the acoustic matching transducer. In Fig. 4.4, the HMM topology simply represents a mapping between each context-dependent phone and a sequence of shared states. All state transitions within a phone are represented in the acoustic matching transducer. The manner of factorization can be chosen according to the balance between generality and efficiency of the decoder. With regard to efficiency, factorization as in Fig. 4.3 or 4.4 is often adopted because the HMM topology handled by a general-purpose program for WFST is smaller than the acoustic matching transducer handled by a dedicated program that usually runs faster than the general-purpose program. By contrast, the factorization as in Fig. 4.2 makes it easier to use any kind of HMM topology in the WFST framework.

Finally, H in Eq. (4.8) is constructed to transduce an HMM-shared-state Id sequence into a context-dependent phone sequence [RPM97]. As a result, the composite WFST $H \circ C \circ L \circ G$ also accepts an HMM-shared-state Id sequence. In [MPR02], H is first constructed as the example of Fig. 4.3(a), and then chained transitions are replaced with one transition after composition and optimization for $H, C, L,$ and G. The input label of the new transition results in the HMM-shared-state Id sequence along the chain as in 4.4(a). This step is described in Section 4.3. In [ARS09], H is not made and the transducer in Fig. 4.1 is handled directly by the decoder program and combined with $CLG = C \circ L \circ G$ during decoding.

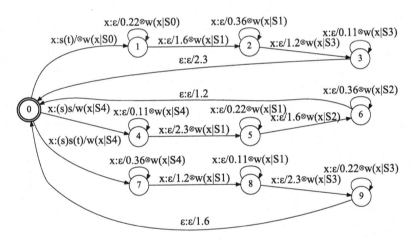

Figure 4.1: HMM transducer

4.2.2 PHONE CONTEXT DEPENDENCY

As described in Section 2.5, subword-based acoustic models are made for context-dependent phone units in most LVCSR systems. Those models have to be connected properly in the search network according to their context dependency. The connections between such context-dependent units are represented as an FST denoted by C.

C is not difficult to construct if the context dependency is limited to one preceding phone and one succeeding phone, i.e., in the case of triphones. A state is prepared for every phone pair and a state transition is made for each triphone. The transition is spread between source and destination

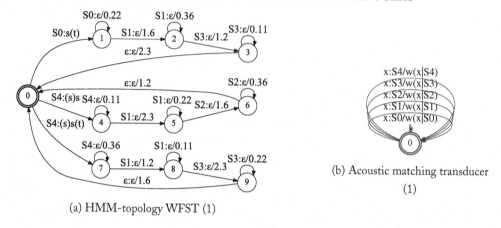

(a) HMM-topology WFST (1)

(b) Acoustic matching transducer (1)

Figure 4.2: Example 1: HMM-topology WFST and acoustic matching transducer

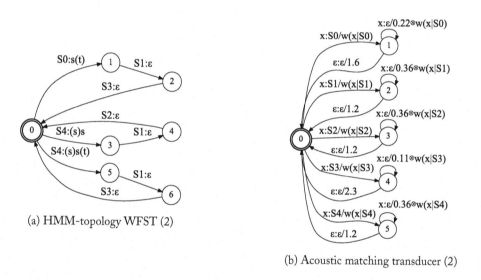

(a) HMM-topology WFST (2)

(b) Acoustic matching transducer (2)

Figure 4.3: Example 2: HMM-topology WFST and acoustic matching transducer

states, where the phone pair of the source state has to match a pair consisting of the left and center phones of the triphone, and the phone pair of the destination state also has to match a pair consisting of the center and right phones.

Figure 4.5 shows WFST C, which represents triphone context dependency when we have just two basic phone units /t/ and /s/. The input label of each transition is a triphone while the output label is a context-independent phone that is equal to the center phone of the triphone. However, the input labels of transitions from the initial state 0 are left-context independent and those to the final state 1 are right-context independent.

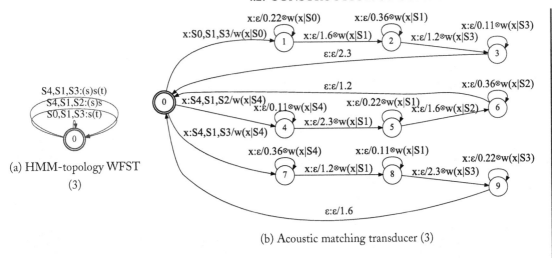

(a) HMM-topology WFST
(3)

(b) Acoustic matching transducer (3)

Figure 4.4: Example 3: HMM-topology WFST and acoustic matching transducer

4.2.3   PRONUNCIATION LEXICON

WFST L, which represents a pronunciation lexicon, can be constructed based on a set of transductions from a phone sequence to a word, where each input label corresponds to a phone unit in the subword-based acoustic models and each output label corresponds to a word in the vocabulary. For continuous speech recognition, L is constructed so that each transduction can be applied repeatedly with cyclic transitions.

Table 4.1 shows an example of a pronunciation, where the words are placed in the left column, and their pronunciations in the right column. Although each word has only one pronunciation in this example, it may have multiple pronunciations with probabilities.

Table 4.1: A simple pronunciation lexicon

word	pronunciation
<s>	sil
</s>	sil
START	s t aa r t
STOP	s t aa p
IT	ih t

Figure 4.6 shows a WFST that represents the example lexicon, where each transduction is represented as a cycle from state 0. By using this WFST, for example, phone sequence "s t aa p ih t" is transduced into word sequence "STOP IT" with state transitions 0, 5, 6, 7, 0, 8, and 0. Since there is only one pronunciation for each word in the lexicon, it is assumed that the probability of pronun-

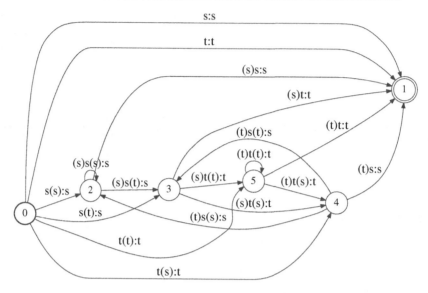

Figure 4.5: Triphone context dependency FST

ciation v for word w, $P(v|w)$, is always one. Hence, the weights are omitted from this figure. But when multiple pronunciations are allowed for each word, the probability of each pronunciation may be placed at the beginning transition from state 0.

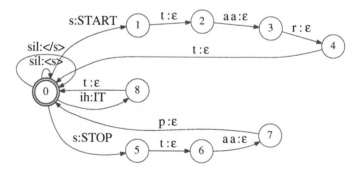

Figure 4.6: Pronunciation lexicon WFST

There is a more flexible way to describe pronunciations in a lexicon, which allows us to use the regular expressions for multiple pronunciations. Since any regular expression can be converted into a finite automaton, we can create L by making a union of automata each representing a single or multiple pronunciation(s) for a word by regular expression and applying a Kleene closure, where an output word label is attached to each corresponding automaton. In this book, we do not describe how to make an automaton from a regular expression (For details see Chapter 3 of [HMU06]).

In addition, we mention dealing with a short pause that can be inserted between any two words. If the language model has a short pause entry, we simply need to add the short pause as a regular word in the lexicon. But if the language model does not have a pause, a special treatment is required for the lexicon WFST. One simple way is to insert a transition with "sp:ε" into L as a self loop at the initial state, where "sp" represents the name of the short pause included in the acoustic model. With this transition, 0 or more repetitions of pause models are allowed between any two words. We may attach a weight to the transition to impose a penalty for the insertion of pause models. If we do not want repetitions of pause models, different types of L can be constructed. For example, we construct it as in Fig. 4.7. However, the above methods sometimes increases the size of the fully composed WFST. Memory-efficient methods for handling short pauses are described in [Gar08]. In addition, techniques for dealing with non-speech events as well as short pauses has been presented in [RSN12].

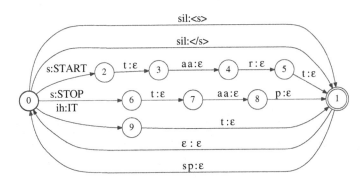

Figure 4.7: Pronunciation lexicon WFST with a short pause

4.2.4 LANGUAGE MODELS

As described in Section 2.6, Finite-State Grammars (FSGs) and n-gram models are widely used as language models for speech recognition. Since FSGs are finite-state models, any FSG or probabilistic FSG (PFSG) can easily be represented as an FSA or a WFSA. Unlike the FSG depicted in 2.8, each word label needs to be assigned to a transition but not a state. Figure 4.8 shows an FSA equivalent to the FSG, which is made by attaching every word label to each corresponding transition.

An n-gram model is equivalent to an $(n-1)$-order Markov model, and therefore can be represented as a WFSA as it is. The $(n-1)$-order Markov model of a language has $|V|^{n-1}$ states and $|V|^n$ transitions, where $|V|$ denotes the vocabulary size. Each state of the model corresponds to an $(n-1)$-word history and each transition labeled with a word corresponds to an occurrence of

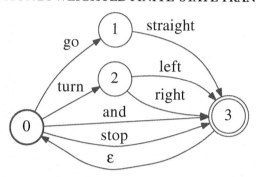

Figure 4.8: Finite-state grammar as an FSA

the word in the history associated with its source state, where an n-gram probability is attached to the transition.

However, since $|V|$ is large and n is 2 or more, e.g., $|V| \geq 50,000$ and $n = 3$ in a standard LVCSR system, an enormous number of states and transitions are necessary for constructing the WFSA, e.g., it requires at least $50,000^2$ states and $50,000^3$ transitions. Thus, it is too expensive for the run-time memory of speech recognition to increase $|V|$ and/or n. This problem is solved by incorporating a back-off mechanism for n-gram probabilities explicitly in the state transitions. As described in Section 2.6.3, back-off smoothing is effective for estimating the probabilities of unseen n-grams using m-gram probabilities such that $m < n$. With this approach, we need probabilities only for the observed n-grams. The probabilities for unseen n-grams are calculated using m-gram probabilities and the back-off coefficients, which are much fewer than those of the observed n-grams.

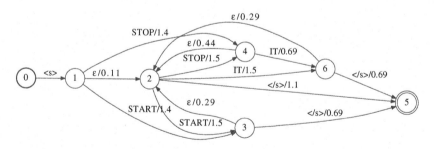

Figure 4.9: Bigram WFSA

Figure 4.9 shows an example of a WFSA representing a back-off bigram model. With a bigram, each state corresponds to a single word history, but one state corresponds to a zero word history, which is prepared for backing off to the unigrams. In the figure, state 2 is of the zero word history. Each of the states except for the initial state 0, zero word history state 2 and the final state 5 is associated with a single word history, where states 1, 3, 4 and 6 correspond to the words "<s>," "START," "STOP," and "IT," respectively. "<s>" and "</s>" are meta symbols that represent the beginning and end of a sentence. In the example WFSA, the bigram probability $P(\text{IT}|\text{STOP})$ is assumed to be 0.5 and its weight is assigned to the state transition 4 to 6 as $-\log P(\text{IT}|\text{STOP}) = 0.69$ in a tropical semiring.

As in the back-off n-gram probability of Eq. (2.18), if a word bigram $w_1 w_2$ is unseen, i.e., $C(w_1, w_2) = 0$, its bigram probability is calculated as $P(w_2|w_1) = \alpha(w_1)P(w_2)$. Suppose we calculate a bigram probability $P(\text{IT}|\text{START})$ using the example WFSA. First we consider a transition from state 3 to 6. But there is no direct transition labeled with "IT" between those states because the bigram "START IT" might be unseen and therefore was not made in the WFSA. Thus, we need to apply back-off by making a transition to state 2 with the back-off coefficient weight $-\log \alpha(\text{START}) = 0.29$. Then, unigram probability weight $-\log P(\text{IT}) = 1.5$ is accumulated by making a transition from 2 to 6. Accordingly, the weight for the bigram probability, $-\log P(\text{IT}|\text{START})$, can be calculated by making state transitions 3, 2, and 6 using the back-off mechanism. For the word sequence "<s> START IT STOP IT </s>," we can calculate the sentence probability with state transitions 0, 1, 3, 2, 6, 2, 4, 6, and 5 by accumulating the weights along the path.

With high-order n-gram models, the back-off mechanism is implemented as well as the bigram model, where each state corresponds to an m-word history, and $0 \le m < n$. If we cannot find any transition labeled with a word in the current state, we can find another transition for the word by making an epsilon transition to the state with the reduced history truncated into $m - 1$ words. This back-off transition can be made iteratively until the word is found or m becomes 0. Note that the back-off mechanism implemented in a WFSA works in a slightly different way from the back-off scheme in Eq. (2.18) because the WFSA is used in a search network for Viterbi decoding, which is composed of other component WFSTs. In a Viterbi algorithm with a WFST, the minimal weighted path is found for a given input sequence when there are different paths that accept the sequence. In fact, we can obtain different paths for a word sequence if we implement the back-off mechanism as in Fig. 4.9. For example, there are two paths for calculating the bigram probability $P(\text{IT}|\text{STOP})$, one makes a transition from state 4 to 6 and the other makes transitions of states 4, 2, and 6. If $P(\text{IT}|\text{STOP}) \ge \alpha(\text{STOP})P(\text{IT})$, the back-off works correctly, otherwise it works incorrectly. Thus, the above implementation is considered an approximate method. The degree of this approximation depends on the discounting method used for back-off smoothing, but it is usually not a big problem because the probability of an observed n-gram is higher than its backed-off probability in most cases. Consequently, this WFSA representation is generally used in WFST-based speech recognition. There is another representation with more states and transitions for implementing the back-off scheme exactly in a tropical semiring [AMR03]. But the difference between the speech recognition accuracies of exact and approximate implementations is usually negligible.

We show an algorithm for constructing a WFSA from an n-gram language model. Algorithm 16 is used for constructing a WFSA that includes an approximate back-off mechanism in a tropical semiring. In the algorithm, we denote each state that corresponds to a history h, as "(h)," where h can be a word sequence, a single word, or a zero word denoted by ε. We use the special symbol (–) just for the initial state of the WFSA.

At the beginning of the algorithm (lines 1-2), there are an initial state (–) and a sentence-begin state (`<s>`), which correspond to states 0 and 1 in Fig. 4.9, respectively. These states are then connected with a transition at line 3. The occurrence probability for `<s>` at the beginning of any sentence is assumed to be one ($\bar{1}$).

In line 4, state (`<s>`) is pushed into queue S, and then the WFSA is constructed in the while loop of lines 5-36. Note that this algorithm works with any queue discipline. In lines 6 and 7, a state is taken from S as a source state, which is identified by the m-word sequence, v_1^m. In lines 8-14, a back-off state (v_2^m) and a transition from state (v_1^m) to (v_2^m) are made unless v_1^m is ε, i.e., a zero-word history. The back-off state is identified as word sequence v_2^m where the first word is truncated from v_1^m. We assume here that v_k^m is ε if $m < k$. The back-off transition is made on line 13 with back-off coefficient $\alpha(v_1^m)$. Through this algorithm, we assume a tropical semiring in which n-gram probabilities and back-off coefficients are converted into weights by the negative logarithm function.

In lines 15-35, a state and a transition are made for the probability of each word w following v_1^m if $P(w|v_1^m)$ is defined in the n-gram model. The procedure in lines 16-22 deals with a special case where w is the sentence end `</s>`. State (`</s>`) is made the final state of the WFSA and a transition is spread from (v_1^m) to the final state.

If w is not the sentence end, state s that corresponds to the destination state from (v_1^m) is prepared in lines 24-32. If m is less than $n - 1$, s is set at (v_1^m, w) or else (v_2^m, w). This means that the length of the word sequence for each state becomes at most $n - 1$. On line 33, a transition from (v_1^m) to s is added with label w and weight $-\log P(w|v_1^m)$. Finally, if queue S becomes empty, the algorithm terminates and returns WFSA G at line 37.

The number of observed n-grams is actually much smaller than $|V|^n$ even in a large-scale corpus, but the number can still be too high when we increase $|V|$ and n. Generally, we can eliminate n-gram probabilities that do not contribute to the speech recognition accuracy. Such probabilities are obtained using the back-off scheme. n-grams to be eliminated from the model can be found as less-counted n-grams or by using an entropy-based pruning technique [Sto98]. With these techniques, we can reduce the size of the n-gram WFSA and also make decoding faster because it also reduces the size of the search network built by the composition of all the components including the n-gram language model.

After constructing an FSA or a WFSA representing a language model, we need to convert it into a WFST that is equivalent to the original acceptor, because we perform the composition operation between the language model and other knowledge sources such as a pronunciation lexicon in WFST form. The WFST can be made easily to have the same output label as the input label in every transition. For example, the transition from state 1 to 3 in Fig. 4.9, "START/1.4," is converted to "START:START/1.4."

Algorithm 16 N-gram-WFSA(V, P, α)

1: $I \leftarrow \{(-)\}$
2: $Q \leftarrow \{(-), (\texttt{<s>})\}$
3: $E \leftarrow \{((-), \texttt{<s>}, \bar{1}, (\texttt{<s>}))\}$
4: $S \leftarrow \{(\texttt{<s>})\}$
5: **while** $S \neq \emptyset$ **do**
6:    $(v_1^m) \leftarrow \text{Head}(S)$
7:    $\text{Dequeue}(S)$
8:    **if** $(v_1^m) \neq \varepsilon$ **then**
9:       **if** $(v_2^m) \notin Q$ **then**
10:          $Q \leftarrow Q \cup \{(v_2^m)\}$
11:          $\text{Enqueue}(S, (v_2^m))$
12:       **end if**
13:       $E \leftarrow E \cup \{((v_1^m), \varepsilon, -\log \alpha(v_1^m), (v_2^m))\}$
14:    **end if**
15:    **for** each $w \in (V - \{\texttt{<s>}\})$ such that $P(w|v_1^m)$ is defined **do**
16:       **if** $w = \texttt{</s>}$ **then**
17:          **if** $(\texttt{</s>}) \notin Q$ **then**
18:             $Q \leftarrow Q \cup \{(\texttt{</s>})\}$
19:             $F \leftarrow \{(\texttt{</s>})\}$
20:             $\rho((\texttt{</s>})) \leftarrow \bar{1}$
21:          **end if**
22:          $E \leftarrow E \cup \{((v_1^m), \texttt{</s>}, -\log P(\texttt{</s>}|v_1^m), (\texttt{</s>}))\}$
23:       **else**
24:          **if** $m = n - 1$ **then**
25:             $s \leftarrow (v_2^m, w)$
26:          **else**
27:             $s \leftarrow (v_1^m, w)$
28:          **end if**
29:          **if** $s \notin Q$ **then**
30:             $Q \leftarrow Q \cup \{s\}$
31:             $\text{Enqueue}(S, s)$
32:          **end if**
33:          $E \leftarrow E \cup \{((v_1^m), w, -\log P(w|v_1^m), s)\}$
34:       **end if**
35:    **end for**
36: **end while**
37: **return** $G = (V, Q, I, F, E, \bar{1}\rho)$

4.3   COMPOSITION AND OPTIMIZATION

We explain the composition and optimization steps for constructing a fully composed WFST according to the method described in [MPR02]. Before composition, we need to modify the component WFSTs H, C, and L to make them determinizable in the following optimization step. We refer to such modified WFSTs as $\tilde{H}$, $\tilde{C}$, and $\tilde{L}$, respectively. Here G is assumed to be deterministic. The n-gram WFST with back-off epsilon transitions is actually deterministic if we regard epsilon to be a regular symbol label. In the determinization operation, we consistently consider epsilons to be regular symbol labels.

Then the composition and optimization are performed step by step for efficient construction as follows:

$$N = \text{fact}(\pi_\varepsilon(\min(\det(\tilde{H} \circ \tilde{C} \circ \det(\tilde{L} \circ G))))), \tag{4.9}$$

where $\det(\cdot)$ and $\min(\cdot)$ denote determinization and minimization operations. $\text{fact}(\cdot)$ and $\pi_\varepsilon(\cdot)$ indicate the factorization and auxiliary symbol removal, which we explain later in this section. Finally, we obtain a fully composed WFST N that can be used for decoding. Once the WFST is constructed, we do not have to reconstruct it until at least one knowledge source is updated.

First we modify the WFST of the pronunciation lexicon, L. The most important thing we need to consider is determinizability. L is often indeterminizable because of the presence of homophonic words. In such a case, L does not become functional, i.e., a phone sequence can be transduced into multiple word sequences using L. Being functional is a sufficient condition for being a determinizable transducer. To make L determinizable, we insert auxiliary symbol labels that distinguish the homophones for different words and make L functional. Since the auxiliary symbol labels are necessary only in determinization, they are replaced with epsilons after the final determinization step.

A transition that has an auxiliary input symbol label is inserted at the end of the transitions for the phone sequence. This is equivalent to extending each pronunciation in the lexicon as follows

```
night    n ay t #1
knight   n ay t #2,
```

where the homophonic words "night" and "knight" are distinguished by auxiliary symbols #1 and #2. If there are M homophonic words, we need to extend their pronunciations with #1, #2, . . . , #M.

Even if we have only one word for a given pronunciation, i.e., there are no other homophonic words, it is better to insert an auxiliary symbol label for the word because a phone sequence of two or more consecutive words might be equal to that of another word or word sequence. For example, "tonight" and "to night" can have the same pronunciation "t ax n ay t." Hence, this transducer cannot be functional. To ensure its functionality, we also need to add #1 for all the non-homophonic words. Consequently, "tonight" and "to night" can be distinguished by their pronunciations "t ax n ay t #1" and "t ax #1 n ay t #1." We denote the lexicon WFST including auxiliary symbol labels as $\tilde{L}$.

Next we combine $\tilde{L}$ and G, and then determinize the resulting WFST as

$$LG = \det(\tilde{L} \circ G). \tag{4.10}$$

If G is the back-off n-gram WFST shown in the previous section, here we need to use a composition operation with an epsilon-sequencing filter to maintain the number of input epsilon transitions in G. Strictly speaking, G is not deterministic, i.e., there can be multiple paths for a given word sequence because of back-off transitions. But if we regard the input epsilons as regular symbols, the WFSA is assumed to be deterministic. Every back-off path can be distinguished from others by the number of epsilons, which corresponds to the number of back-offs along the path. Hence, to determinize $\tilde{L} \circ G$, those epsilons must not be removed before performing the composition, or else we cannot ensure that the $\tilde{L} \circ G$ result is determinizable, even if pronunciations are distinguishable from each other in $\tilde{L}$. However, since the epsilon-matching filter consumes an input epsilon of G and an output epsilon of $\tilde{L}$ simultaneously, the original number of input epsilons in G is not maintained. Therefore, those back-off paths cannot be distinguished from each other. On the other hand, the epsilon-sequencing filter keeps G's epsilon transitions in $\tilde{L} \circ G$.

We show examples of $L \circ G$ and $\det(L \circ G)$ in Figs. 4.10 and 4.11, respectively, where we omit auxiliary symbols because of space limitations. The determinized WFST $\det(L \circ G)$ is smaller than that of $L \circ G$, and there is at most one transition for an input label from each state, i.e., the WFST is deterministic.

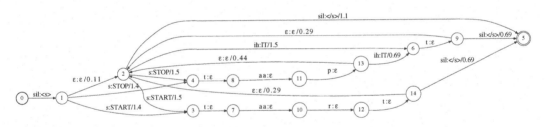

Figure 4.10: $L \circ G$

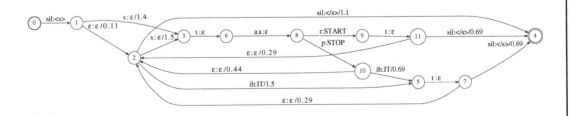

Figure 4.11: $\det(L \circ G)$

After the construction of LG, we combine it with C. But, first determinizing C with respect to output labels is important. We can obtain such a WFST C' with invert(det(invert(C))), where invert($\cdot$) indicates the inversion of the WFST. Figure 4.12 shows C' when C is the triphone context-dependency FST in Fig. 4.5.

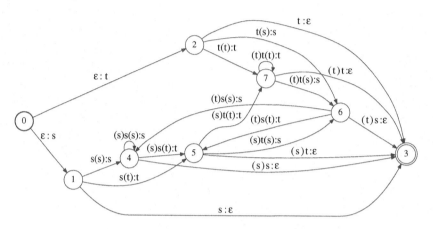

Figure 4.12: $C' = $ invert(det(invert(C)))

Since C' is deterministic with respect to output labels, the composition of C' and LG does not significantly exceed the size of LG. In addition, C' is almost deterministic with respect to input labels except for epsilon transitions from the initial state as shown in Fig. 4.12. Therefore, $C' \circ LG$ also results in an almost deterministic WFST. The composition of LG in Fig. 4.11 and some C' actually becomes a deterministic WFST because LG has only one transition from the initial state. Hence, it is usually unnecessary to determinize the WFST for $C' \circ LG$.

Moreover, we must not forget to insert auxiliary symbols into C' before the composition of C' and LG. LG still has auxiliary symbol labels that are necessary for the succeeding optimization steps. But since C' does not have such output labels, transitions with auxiliary symbol labels will be lost in the composition operation. Accordingly, we need to add self loop transitions with those labels to every state in C' so that any auxiliary symbol can be inserted in any word pronunciation. If we know the maximum number of homophonic words M in the lexicon, M self-loop transitions with the labels #1, #2, ..., #M are added to each state of C'. Such transitions should have the same input and output labels as "#m :#m." We refer to this modified WFST as $\tilde{C}$. With this modification, the result of

$$CLG = \tilde{C} \circ LG \tag{4.11}$$

still has the auxiliary symbol labels, which keep the WFST determinizable.

In the next step, H is modified by adding self loops with auxiliary symbol labels to the initial state, where we assume that the initial state is located at the boundary between any phones as in Figs. 4.2–4.4. As with C, we insert M transitions into the initial state. We refer to this modified H

as $\tilde{H}$. The final composition and determinization are performed as

$$HCLG = \det(\tilde{H} \circ CLG). \qquad (4.12)$$

Then $HCLG$ is minimized, and the auxiliary symbols are removed as follows

$$HCLG' = \pi_\varepsilon(\min(HCLG)), \qquad (4.13)$$

where $\pi_\varepsilon(\cdot)$ replaces all the auxiliary symbol labels with ε.

Finally, factorization can be applied to $HCLG'$ as

$$N = \mathrm{fact}(HCLG'). \qquad (4.14)$$

The factorization is effective in terms of WFST size. The numbers of states and transitions of $HCLG'$ can be reduced by replacing chained transitions with a single transition, where the labels on the chain are concatenated and the weights are cumulated with $\otimes$-multiplication. Figure 4.13 shows an example of factorization, where (a) is the original and (b) is the result. As shown in the

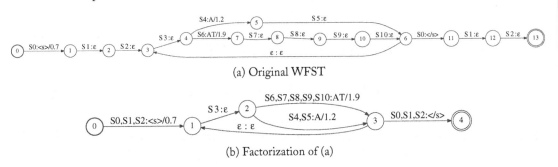

(a) Original WFST

(b) Factorization of (a)

Figure 4.13: Example of factorization

figure, each transition no longer has a single input label, i.e., it is a sequence of HMM-shared states like the example in Fig. 4.4(a), where a unique shared-state sequence on a transition is assumed to be a sub-word HMM. State transitions within such a sub-word HMM need to be simulated in a decoder based on a transducer as in the example shown in Fig. 4.4(b). Note that each shared-state sequence in Fig. 4.13(b) has different lengths as a result of factorization unlike Fig. 4.4(a).

In summary, we show the size of each WFST generated according to the above steps in Table 4.2. We constructed a pronunciation lexicon, acoustic and language models with the Corpus of Spontaneous Japanese (CSJ) [FMI00], which includes about 2,000 lecture speeches of 10 to 30 minutes' duration. The acoustic model had 5,000 shared states, each of which had a unique probabilistic density function of a Gaussian mixture. The pronunciation lexicon covered a 100k-word vocabulary. The language model was a back-off trigram model estimated with the transcripts of those lectures. As shown in Table 4.2, the fully composed WFST N is approximately 1.5 times larger than the language model WFST G, i.e., N does not result in an enormous WFST. This is almost the same tendency as that reported in the original work [MPR02].

Table 4.2: The sizes of WFSTs in 100k-word vocabulary CSJ task

	#state	#transition
H	6,518	26,634
C	1,894	85,185
L	501,759	602,566
G	220,773	1,177,625
$\tilde{L} \circ G$	1,510,293	2,849,833
$LG = \det(\tilde{L} \circ G)$	1,184,192	2,228,092
$CLG = \tilde{C} \circ LG$	1,195,569	2,318,779
$\tilde{H} \circ CLG$	3,145,430	4,268,640
$HCLG = \det(\tilde{H} \circ CLG)$	3,034,175	4,102,668
$HCLG' = \pi_\varepsilon(\min(HCLG))$	2,698,861	3,667,087
$N = \text{fact}(HCLG')$	736,221	1,704,447

In the following sections, we describe a decoding algorithm with a single WFST for speech recognition, and then show evaluation results on decoding performance in the CSJ task.

4.4   DECODING ALGORITHM USING A SINGLE WFST

In this section, we present a speech recognition algorithm using a single fully composed WFST, which is based on the time-synchronous Viterbi beam search. Although our WFST-based algorithm is similar to that for conventional speech recognition given as Algorithm 1, it is generalized for handling the WFST. Algorithm 17 shows a pseudo code of the main procedure, which calls sub-codes for initialization, epsilon transitions, regular (non-epsilon) transitions, and finalization for input feature vector sequence $X = x[1], \ldots, x[T]$. The pseudo code finally generates the recognition result by backtracking the minimally weighted path.

In the sub-codes, we use the following quantities, which are similar to those used in the conventional method.

$\alpha(t, s)$: Cumulative weight of a partial path up to time frame t at a WFST state s, which equals the negative of the Viterbi score in a tropical semiring.

$B(t, s)$: Back pointer to keep track of the most likely path up to time frame t at a WFST state s. $B(t, s)$ takes a pair $\langle \tau, e \rangle$, where τ indicates the starting frame of the transition e, and e is the most likely transition incoming to state s at time frame t. Let $e = 0$ if there is no transition incoming to state s.

$\alpha(t, e, j)$: Cumulative weight of a partial path up to time frame t at an HMM state j in transition e.

Algorithm 17 Single-WFST-ViterbiBeamSearch($N = (\Sigma, \Delta, Q, I, F, E, \lambda, \rho)$, $X = x[1], \ldots, x[T]$)

1: $S \leftarrow$ initialize(I, λ)
2: $A \leftarrow \emptyset$
3: **for** $t = 1$ to T **do**
4:     $S \leftarrow$ transition_with_epsilon($E, S, t - 1$)
5:     $\langle S, A \rangle \leftarrow$ transition_with_input($E, S, A, x[t], t$)
6:     prune(S, A, t)
7: **end for**
8: $\hat{B} \leftarrow$ final_transition(E, F, ρ, S, T)
9: $Y \leftarrow$ backtrack($\hat{B}$)
10: **return** Y

$b(t, e, j)$: Back pointer to keep track of the most likely path up to time frame t at an HMM state j in transition e. $b(t, e, j)$ holds only the starting frame of the transition e.

In the following pseudo codes, the input label of each transition can be an HMM state sequence, which is considered a subword HMM but does not necessarily correspond to a triphone model. For example, input labels "S0,S1,S2," "S3," "S4,S5," and "S6,S7,S8,S9,S10" in Fig. 4.13 are considered subword HMMs. We assume that the subword HMM for each transition e has an initial state i_e and a final state f_e in addition to regular HMM states each of which has a probability density function. We also introduce an acoustic weight function $\omega(x, k | \mathcal{M}, j)$ to obtain the transition weight from state j to k of subword HMM $\mathcal{M}$ with feature vector x, which is calculated as

$$\omega(x, k | \mathcal{M}, j) = \begin{cases} -\log a_{jk}^{(\mathcal{M})} b_k^{(\mathcal{M})}(x) & \text{if } x \neq \varepsilon \\ -\log a_{jk}^{(\mathcal{M})} & \text{if } x = \varepsilon \end{cases} \tag{4.15}$$

in a tropical semiring.

The Viterbi search starts from initialization. The pseudo code, initialize(I, λ), is shown in Algorithm 18. In the code, a set of initial states, I, and the initial weight function λ are used to generate initial hypotheses at time frame 0, each of which has a weight $\alpha(0, i) = \lambda(i)$ for state i in I. The back pointers $B(0, i)$ are all set at $\langle 0, 0 \rangle$. Then all the initial states are inserted in queue S to extend the hypotheses from those states in the following steps.

For each time frame t, epsilon transitions are first made, and then regular transitions are made with the input vector $x[t]$. The pseudo code for epsilon transitions, transition_with_epsilon(E, S, t), is shown in Algorithm 19. In lines 5–14, epsilon transitions are made from each state in queue S, and new hypotheses for the destination states are made with their weights and back pointers. We use $\oplus$-addition to compare different weights as line 7, where $\alpha \oplus \alpha' = \alpha'$ means α' is better than α in an idempotent semiring such as a tropical semiring. This is substituted with $\min(\alpha, \alpha') = \alpha'$ in a tropical semiring. In line 11, the destination state $n[e]$ is pushed into queue S if $n[e]$ is not in S,

---

Algorithm 18 initialize(I, λ)

1: **for** each $i \in I$ **do**
2:     $\alpha(0, i) \leftarrow \lambda(i)$
3:     $B(0, i) \leftarrow \langle 0, 0 \rangle$
4:     Enqueue(S, i)
5:     **return** S
6: **end for**

---

since epsilon transitions can be made repeatedly. In lines 15-17, each state s in S is inserted into S' if there is at least one regular transition from state s. S' holds active states from which the subsequent regular transitions will be made.

Algorithm 19 transition_with_epsilon(E, S, t)

1: $S' \leftarrow \emptyset$
2: **while** $S \neq \emptyset$ **do**
3:     $s \leftarrow$ Head(S)
4:     Dequeue(S)
5:     **for** each $e \in E(s, \varepsilon)$ **do**
6:         $\alpha' \leftarrow \alpha(t, s) \otimes w[e]$
7:         **if** $\alpha(t, n[e]) \oplus \alpha' = \alpha'$ **then**
8:             $\alpha(t, n[e]) \leftarrow \alpha'$
9:             $B(t, n[e]) \leftarrow \langle t, e \rangle$
10:             **if** $n[e] \notin S$ **then**
11:                 Enqueue($S, n[e]$)
12:             **end if**
13:         **end if**
14:     **end for**
15:     **if** $s \notin S'$ and $\{e | e \in E(s), i[e] \neq \varepsilon\} \neq \emptyset$ **then**
16:         Enqueue(S', s)
17:     **end if**
18: **end while**
19: **return** S'

The pseudo code for regular transitions, transition_with_input(E, S, A, x, t), is shown in Algorithm 20. In lines 1-9, the hypotheses go into the HMM states of each transition outgoing from each state s in S. For i_e, the initial state of the HMM of transition e, cumulative weight $\alpha(t - 1, e, i_e)$ and back pointer $b(t - 1, e, i_e)$ are assigned in lines 3 and 4. The initial HMM states are inserted in queue A together with the transition in line 6.

Algorithm 20 transition_with_input(E, S, A, x, t)

1: **for** each $s \in S$ **do**
2:   **for** each $e \in E(s)$ such that $i[e] \neq \varepsilon$ **do**
3:     $\alpha(t - 1, e, i_e) \leftarrow \alpha(t - 1, s) \otimes w[e]$
4:     $b(t - 1, e, i_e) \leftarrow t - 1$
5:     **if** $\langle e, i_e \rangle \notin A$ **then**
6:       $A \leftarrow A \cup \{\langle e, i_e \rangle\}$
7:     **end if**
8:   **end for**
9: **end for**
10: $S' \leftarrow A' \leftarrow \emptyset$
11: **while** $A \neq \emptyset$ **do**
12:   $\langle e, j \rangle \leftarrow \text{Head}(A)$
13:   $\text{Dequeue}(A)$
14:   **for** each $k \in \text{Adj}(j)$ such that $k \neq f_e$ **do**
15:     $\alpha' \leftarrow \alpha(t - 1, e, j) \otimes \omega(x, k | i[e], j)$
16:     **if** $\alpha(t, e, k) \otimes \alpha' = \alpha'$ **then**
17:       $\alpha(t, e, k) \leftarrow \alpha'$
18:       $b(t, e, k) \leftarrow b(t - 1, e, j)$
19:       **if** $\langle e, k \rangle \notin A'$ **then**
20:         $\text{Enqueue}(A', \langle e, k \rangle)$
21:       **end if**
22:     **end if**
23:   **end for**
24: **end while**
25: **for** each $\langle e, k \rangle \in A'$ such that $f_e \in \text{Adj}(k)$ **do**
26:   $\alpha' \leftarrow \alpha(t, e, k) \otimes \omega(\varepsilon, f_e | i[e], k)$
27:   **if** $\alpha(t, n[e]) \oplus \alpha' = \alpha'$ **then**
28:     $\alpha(t, n[e]) \leftarrow \alpha'$
29:     $B(t, n[e]) \leftarrow \langle b(t, e, k), e \rangle$
30:     **if** $n[e] \notin S'$ **then**
31:       $\text{Enqueue}(S', n[e])$
32:     **end if**
33:   **end if**
34: **end for**
35: **return** $\langle A', S' \rangle$

Algorithm 21 prune(S, A, t)

1: $w_t^{best} \leftarrow \left\{ \bigoplus_{s \in S} \alpha(t, s) \right\} \oplus \left\{ \bigoplus_{\langle e, j \rangle \in A} \alpha(t, e, j) \right\}$

2: $w_t^{th} \leftarrow \gamma \otimes w_t^{best}$

3: **for** each $s \in S$ **do**

4:    **if** $\alpha(t, s) \oplus w_t^{th} = w_t^{th}$ **then**

5:       Erase(S, s)

6:    **end if**

7: **end for**

8: **for** each $\langle e, j \rangle \in A$ **do**

9:    **if** $\alpha(t, e, j) \oplus w_t^{th} = w_t^{th}$ **then**

10:       Erase($A, \langle e, j \rangle$)

11:    **end if**

12: **end for**

13: **return** $\langle S, A \rangle$

HMM-level state transitions are made in lines 10-24. For each pair $\langle e, j \rangle$ in A, HMM-level transitions are made from state j in the HMM associated with transition e. In line 14, Adj(j) returns the adjacency list for state j, i.e., the set of HMM states that are reachable by one transition from state j in the HMM. For each transition from state j to k, the cumulative weight and the back pointer are calculated with the acoustic distance $D(i[e], j, k, x)$, and the best cumulative weight and the back pointer at the destination state k are recorded in $\alpha(t, e, k)$ and $b(t, e, k)$, respectively. These new active HMM states are stored in queue A' in line 20.

In lines 25-34, the cumulative weight and the back pointer outgoing from each transition e at frame t are given to the next state $n[e]$ as $\alpha(t, n[e])$ and $B(t, n[e])$. State $n[e]$ is also stored in S' for further WFST-level transitions.

Finally, in line 35, active HMM states and WFST states at frame t are returned to the main code.

After the epsilon and regular transitions over T frames are complete, final epsilon transitions are made with Algorithm 22. In lines 3-14, epsilon transitions are made from active states in S at the final frame T. For all the active final states, the best cumulative weight $\hat{\alpha}$ and the back pointer $\hat{B}$ are obtained through lines 15-21.

Using the backtracking of Algorithm 23, the output label sequence on the best path is picked up for the recognition result. The sequence is obtained by back tracking the path from the best back pointer $\hat{B}$ as in lines 2-6.

Thus, a Viterbi search using a single WFST can be performed in a similar way to the conventional decoding algorithm (Algorithm 1). As with the conventional approaches, the decoding algorithm for WFSTs can also be extended to generate lattices by keeping multiple back pointers [LPR99].

Algorithm 22 final_transition(E, F, ρ, S, T)

1: $\hat{\alpha} \leftarrow \bar{0}$
2: **while** $S \neq \emptyset$ **do**
3:    $s \leftarrow \text{Head}(S)$
4:    $\text{Dequeue}(S)$
5:    **for** each $e \in E(s, \varepsilon)$ **do**
6:      $\alpha' \leftarrow \alpha(T, s) \otimes w[e]$
7:      **if** $\alpha(T, n[e]) \oplus \alpha' = \alpha'$ **then**
8:        $\alpha(T, n[e]) \leftarrow \alpha'$
9:        $B(T, n[e]) \leftarrow \langle T, e \rangle$
10:       **if** $n[e] \notin S$ **then**
11:          $\text{Enqueue}(S, n[e])$
12:       **end if**
13:      **end if**
14:    **end for**
15:    **if** $s \in F$ **then**
16:      $\alpha' \leftarrow \alpha(T, s) \otimes \rho(s)$
17:      **if** $\hat{\alpha} \oplus \alpha' = \alpha'$ **then**
18:        $\hat{\alpha} \leftarrow \alpha'$
19:        $\hat{B} \leftarrow B(T, s)$
20:      **end if**
21:    **end if**
22: **end while**
23: **return** $\hat{B}$

4.5   DECODING PERFORMANCE

Finally, we mention the decoding performance of WFST-based speech recognition. The decoding performance in speech recognition is evaluated with a pair of the recognition accuracy and the decoding time, where the performance is assumed to be high if the decoder yields a high recognition accuracy and a short decoding time at the same time. In a Viterbi beam search, the accuracy and the decoding time change depending on the beam width. The time becomes shorter for a narrower beam, but the accuracy decreases due to increase of pruning errors. If we use a wider beam, the accuracy comes close to the maximum that will be obtained when we turn off beam pruning. In this section, we show the impact of WFST optimization on the decoding performance in CSJ task.

    Figure 4.14 shows relationships between word accuracy and real time factor when using different WFSTs. The word accuracy is a measure of recognition accuracy, which is calculated as

$$WACC[\%] = \frac{NW - SUB - INS - DEL}{NW} \times 100, \tag{4.16}$$

Algorithm 23 backtrack($\hat{B}$)

1: $\hat{W} \leftarrow \varepsilon$
2: $\langle \hat{t}, \hat{e} \rangle \leftarrow \hat{B}$
3: **while** $\langle \hat{t}, \hat{e} \rangle \neq \langle 0, 0 \rangle$ **do**
4: $\quad \hat{W} \leftarrow o[\hat{e}] \cdot \hat{W}$
5: $\quad \langle \hat{t}, \hat{e} \rangle \leftarrow B(\hat{t}, p[\hat{e}])$
6: **end while**
7: **return** $\hat{W}$

---

where NW denotes the number of actually spoken words, SUB, INS, and DEL denote the numbers of substitution, insertion, and deletion errors, respectively. These numbers can be obtained by aligning the correct word sequence (reference) and the recognized word sequence (hypothesis) using a dynamic programming technique. The real time factor indicates a normalized recognition time, which is obtained as the ratio of decoding time to utterance time.

In Fig. 4.14, each line corresponds to decoding performances for a WFST when changing the beam width. The first WFST was fully optimized according to Eq. (4.9), which was the same as N in Table 4.2. The line is specified as "Optimized" in the figure. The second WFST was optimized but combined with the original C that was not determinized by output labels. The line is specified as "Optimized except C." The third WFST was not optimized, whose line is specified as "Not optimized." We measured the recognition time on a multi-core Xeon X5570 3GHz processor when the decoder ran as a single process. These results show that optimization of the fully composed WFST has a big impact on the decoding performance in speech recognition. The result for "Optimized except C" also shows that determining C by output labels is important in the construction process.

In addition, we examine what semiring should be used through the optimization steps. This choice may affect the decoding performance of the Viterbi beam search. Actually, the semiring affects the distribution of weights over each path in the WFST as a result of determinization or weight pushing.

In determinization, each weight is pushed toward the succeeding transitions when binding the states and transitions. This corresponds to the $\oplus$-sum at line 9 in Algorithm 11. If we construct component WFSTs as described in this chapter, weights in N will be well distributed as a result of determinization. Before determinizing $(\tilde{L} \circ G)$, most weights are placed at the beginning of intra-word transitions. Such weights are gradually pushed toward the succeeding transitions during the determinization operation.

In weight pushing, each weight is pushed to the initial state. When using a tropical semiring, the weights tend to be located as early as possible on each path. In log semiring, the weights tend to be distributed more evenly than those in a tropical semiring. The earlier application of weights is somewhat effective for pruning unpromising hypotheses during the Viterbi beam search, but increases the risk of pruning errors. The choice of semiring depends on the task.

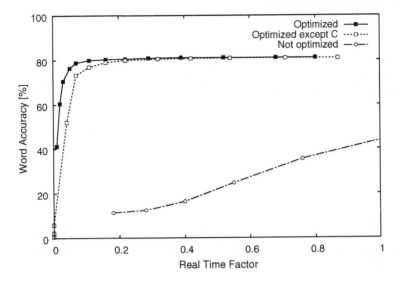

Figure 4.14: Decoding performance with and without optimization in CSJ task.

Moreover, we confirm whether or not weight pushing should be performed as a preprocessing step for WFST minimization. If we apply weight pushing to $HCLG$ before minimization, the weights can be pushed beyond one word. In determinization, the weights just move within one word because the binding procedure necessarily stops within one word due to the auxiliary symbols. A recent study reported that an optimization procedure including determinization in log semiring and excluding weight pushing from minimization is effective for decoding. A case only using CLG without minimization is also used [ARS09].

Figure 4.15 shows the recognition performances for WFSTs optimized in log and tropical semirings with and without weight pushing in the CSJ task. From these results, it is shown that a log semiring is better than a tropical semiring in WFST optimization. This tendency is the same as that reported in [MR01]. In addition, we can see that weight pushing is not necessarily required in WFST minimization by comparing the results with and without weight pushing, i.e., we do not have to push weights beyond one word. However, this could not be true for all WFSTs. If L and G are made unlike the examples in Figs. 4.6 and 4.9, the weight pushing might be necessary to obtain the best performance.

Thus, we need to be careful of what types of component WFSTs are used when combining and optimizing them. If we have more time for building a fully composed WFST, a generalized construction procedure [AMRR04] can be used for combining and optimizing many types of component WFSTs for speech recognition.

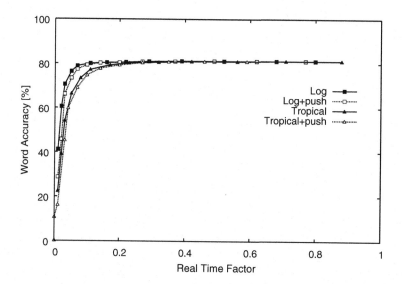

Figure 4.15: Decoding performance when optimizing the WFST in log and tropical semirings with and without weight pushing in CSJ task.

CHAPTER 5

Dynamic Decoders with On-the-fly WFST Operations

As we mentioned in Chapter 4, the WFST approach enables us to accelerate the decoding process for speech recognition. Moreover, the speech recognition network construction process, which is performed by a series of WFST operations using knowledge sources consistently represented in WFST form, is more general than conventional methods. This means that it provides a high modularity for designing a speech recognition network, and thus allows the system to be greatly expanded.

However, we need to consider two problems in real use that do not arise with the conventional methods. One is the large memory consumption by the recognition network, which is usually a large static network in WFST form. The other is the computational cost for the online manipulation of composed and optimized WFSTs. In other words, we have to reconstruct the entire recognition network when we modify a knowledge source even if it is a minor revision, e.g., adding a new word to the vocabulary. To address these problems, the native WFST approach has been extended by the use of a dynamic network that is constructed during decoding with on-the-fly WFST operations. This extension reduces the run-time memory consumption, and also makes the online manipulation easier although it needs a certain overhead because of the on-the-fly operations. In this chapter, we present two dynamic decoding approaches, *look-ahead composition* and *on-the-fly rescoring*, which are more flexible for real applications,[1] as solutions of the above problems.

5.1 PROBLEMS IN THE NATIVE WFST APPROACH

Here we describe the problems with the native WFST approach in more detail. Actually, a fully composed WFST for large-vocabulary continuous speech recognition (LVCSR) often becomes oversized since the full WFST network is made of the multiplicative combination of states and transitions of the component WFSTs. In conventional approaches to LVCSR, such a large network is not usually constructed fully in the memory because its size exceeds the upper bound of the memory size of standard computers. Therefore, the network is dynamically constructed on demand during decoding, but a large amount of the decoding process is devoted to this dynamic construction.

In the WFST approach, the use of the static network is made possible due to the optimization operations. In addition, the efficient language model representation including back-offs (as described in Section 4.2.4) also helps to limit the network to a practical size. Accordingly, the overhead

---

[1] See results in [DHK12], which experimentally compares these decoding methods in terms of speed and memory consumption. We briefly mention this comparison in the last section of this chapter.

consumed for constructing the dynamic network with the conventional approaches has been excluded completely from the decoding process. This is the main reason for the fast decoding. However, the memory consumption is still much larger than those of the conventional approaches even if the network is optimized successfully. In bad cases, the fully composed WFST can easily bloat depending on the specifications of the component WFSTs.

Mohri et al. reported that the total number of transitions of the WFST was 1.4 times that of the trigram language model for a 40k-word vocabulary task [MPR02]. This does not seem to be a serious problem in terms of memory consumption. But the language model is typically the largest component in speech recognition. In addition, WFST representation requires a larger memory than that for the dedicated data structure used in the conventional approaches for language models. For instance, the run-time memory for the authors' WFST-based system using a trigram language model became two to four times larger than that of the conventional systems with the same models. Furthermore, when we used a class-based n-gram model, the size increased several times compared with when we used a model that was not class based [HN05]. Thus, people who handle a WFST-based ASR system may have to pay attention to its memory consumption.

Another problem with the WFST approach is that it offers less flexibility as regards knowledge updates. For example, the lexicon, and the language model need to be updated when we add new words or adapt the system to a new topic. However, once a single WFST is composed and optimized, such partial knowledge updates cannot be accomplished easily. Whenever changes are made to any of the knowledge sources, it is necessary to update the related component WFSTs. Therefore, it requires a long time and usually a large memory to reconstruct a fully composed and optimized WFST. Hence, the WFST approach is considered to be unsuitable for some applications in which the user frequently modifies the knowledge sources of the ASR system.

On the other hand, a multi-pass search strategy is often chosen in order to reduce computation and memory use [ONA97]. In a general two-pass search method, the decoder uses a less complex language model (usually a bigram model) in the first pass, and generates a word lattice including multiple hypotheses. In the second pass, the decoder rescores the lattice using a more complex language model (usually a trigram or a higher-order n-gram model), and selects a better hypothesis in the lattice. This method can also work in the WFST framework [LPR99]. However, the multi-pass approach has a drawback for on-line applications. It has a certain latency after an utterance. Although the first pass can be performed time-synchronously with speech input, the second pass has to await the completion of the first pass, which cannot be completed until the end point of the utterance is detected. In addition, since a certain amount of computation and memory is still necessary for the construction of a fully composed WFST even with a bigram language model, the difficulty involved in the online manipulation of the knowledge sources is not completely excluded by the multi-pass approach.

5.2   ON-THE-FLY COMPOSITION AND OPTIMIZATION

A practical alternative to using a fully composed WFST is on-the-fly composition (sometimes called on-demand, lazy, or delayed composition) of separated WFSTs [MR97, MPR02, DH01, WMMK01, CT01, CT03, HHM04, HN05, HHMN07, CDD07, MSK07, ODIF09a, ARS09]. In this approach, component WFSTs are divided into some groups and one WFST is composed and optimized for each group. A one-pass decoder partially combines these WFSTs during decoding as necessary.

For example, a set of WFSTs for speech recognition, $\{H, C, L, G\}$, is divided into $\{H, C, L\}$ and $\{G\}$, where H, C, L, and G are component WFSTs of HMMs, triphone context dependency, a pronunciation lexicon, and an n-gram language model, respectively. Next HCL is composed as

$$HCL = fact(\pi_\varepsilon(min(det(\tilde{H} \circ (\tilde{C} \circ det(\tilde{L})))))), \tag{5.1}$$

in a similar way to the process used for constructing a fully composed WFST N given in Eq. (4.9) of Section 4.3. HCL construction is computationally much cheaper than that of N, because the largest component G is not included, i.e., determinization or minimization is applied to a smaller WFST at each step. Since G is already deterministic if G is made from an n-gram model, the optimization of G can be omitted.

During decoding, the two WFSTs HCL and G are combined on the fly. Since HCL and G are smaller than N, we can save the run-time memory of the decoder. Each state or transition of $HCL \circ G$ is composed for the first time when it is required in the Viterbi decoding. Once the state or transition is made, it can be kept in the memory and therefore does not need to be composed when it is required again. If we want to save the memory, we may delete some of them that are not frequently used, and compose them again as necessary.

The on-the-fly approach is based on the fact that most WFST operations can be performed on the fly. When we perform an on-the-fly operation with WFST(s), at first no states or transitions of the resulting WFST exist except for the initial state(s). They are made on demand when the resulting WFST is used by another operation or an application such as a speech recognition decoder. The operations, namely union, concatenation, closure, composition, projection, inversion, determinization, epsilon removal, etc., can be used on the fly.

The algorithms of such operations basically consist of the following steps:

1. generate initial states and place them in a queue.

2. repeat steps (a)-(c) while the queue is not empty.

   (a) remove a state from the queue according to an arbitrary queue discipline,

   (b) generate transitions outgoing from the state and their destination states according to the operation if they have not been generated yet, and

   (c) place each new destination state in the queue.

With the above steps, we can generate states and transitions in order of traversing the WFST from the initial states, where we can choose any traversing order including depth-first, breadth-first, and best-first orders. The order is determined by the queue discipline to be used. This type of algorithm can be extended easily to its on-the-fly version.

Suppose a set of transitions is required that are outgoing from an already generated state. If the state is included in the queue, this means that the transitions outgoing from this state have not been generated yet. Therefore, the state is taken out of the queue and the new transitions and their destination states are generated according to steps 2.(a)-(c). If the state is not included in the queue, this means that the transitions are already generated. Such transitions are simply returned. In this way, we can use states and transitions that are generated on demand, but it is required that their source states have ever been used.

5.3 KNOWN PROBLEMS OF ON-THE-FLY COMPOSITION APPROACH

We mentioned that decoding with on-the-fly composition was economical in terms of memory, but was slower than that with a fully composed static WFST. One reason for this slowdown is obviously the computational overhead required for on-the-fly composition. Another reason is that a WFST composed on the fly is insufficiently optimized. To solve the latter problem, on-the-fly determinization may be available to further optimize the composite WFST during decoding. However, the determinization is not actually helpful because the composite WFST is already subsequential (i.e., deterministic for input symbols) according to the fact that the composition of subsequential WFSTs results in a subsequential WFST, where HCL and G are both subsequential. In reality, the WFST composed on the fly is not optimal as regards two factors, the existence of dead states and inefficient weight distributions along the paths. Although trimming and weight pushing operations are usually used to optimize a WFST with respect to these two factors, neither operations can be performed on the fly. We provide more detail about why this problem occurs in the on-the-fly composition.

Figure 5.1 shows examples of WFSTs HCL and G. Suppose these two WFSTs are combined on the fly during decoding. As shown in this example, each arc of HCL has an input label representing a sequence of HMM-state Id numbers and an output label representing an epsilon or a word such as A, B, and C. More epsilons appear on the output side because more Id numbers are necessary to represent a time series of acoustic patterns for a word. On the other hand, each arc of G has only a word label or an epsilon, which is weighted with an n-gram probability or a back-off coefficient. We need to use a filter WFST in the composition operation to deal with epsilon transitions, but omit it from the following explanation for the sake of simplicity.

First, an initial state $(0, 0)$ is composed by coupling the initial states of HCL and G. Once we start decoding, the succeeding states are gradually composed. The next states $(1, 0)$ and $(5, 0)$ are composed to make transitions from the initial state $(0, 0)$. Since the state transitions 0 to 1 and 0 to 5 in HCL do not output any symbols, i.e., they are labeled with ε, G does not make any transitions and stays at state 0.

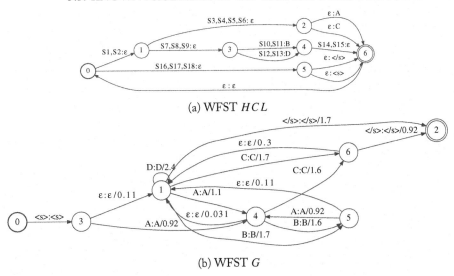

(a) WFST HCL

(b) WFST G

Figure 5.1: Example WFSTs HCL and G to be combined on the fly

Figure 5.2 shows the progress of an on-the-fly composition, where each path from the initial state has become capable of reaching states $(2, 0)$, $(3, 0)$, and $(5, 0)$. However, the transitions from

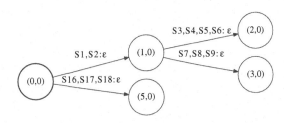

Figure 5.2: Example of on-the-fly composition (1)

states $(2, 0)$ and $(3, 0)$ are not composed and the paths cannot go to the next states, because the transition from state 0 in G only accepts `<s>` and there is no transition that outputs `<s>` from states 2 and 3 in HCL. Consequently, states $(2, 0)$ and $(3, 0)$ become dead states that are not accessible to any final states. State $(1, 0)$ is also a dead state since the paths through state $(1, 0)$ only go to the dead states.

The problem with dead states in on-the-fly composition is that the decoder has to devote a certain amount of computation to dead states and transitions, even though the computation does not contribute at all to the search for the best successful path. In the example in Fig. 5.2, only one transition from state $(5, 0)$ can be composed and this makes it possible to go to the next state.

The occurrence of such dead states and transitions is caused by consecutive output epsilons of the first WFST and/or consecutive input epsilons of the second WFST. With HCL and G, the composition operation repeats to compose states and transitions with output epsilons along each path in HCL until it meets an output word label, while states in G do not change for these epsilons. Since we do not know whether the composed states and transitions are dead or not during this process, we need to compose these states and transitions even though they are not accessible to any final states.

Figure 5.3 shows a WFST after further progress in on-the-fly composition from Fig. 5.2. The

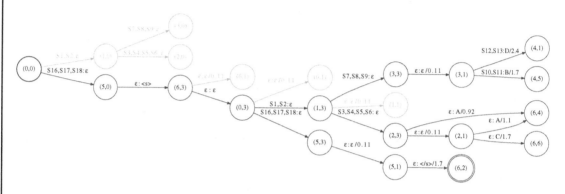

Figure 5.3: Example of on-the-fly composition (2)

nodes and arcs shown in gray represent dead states and transitions that are generated mainly in the paths from the initial state $(0, 0)$ to states $(2, 0)$ and $(3, 0)$. The other dead states and transitions such as $(6, 3) - (6, 1)$ and $(0, 3) - (0, 1)$ are generated for input and output epsilons by using the 2-state filter in Section 3.5. Thus, it is not avoidable to generate dead states and transitions for some WFSTs during the composition operation.

It is possible to generate many dead states and transitions by composition of HCL and G. HCL usually has many consecutive output epsilons, and most word labels are placed after these epsilons. This is a result of determinization for L, in which most output labels of L are moved to the succeeding transitions. In this case, states and transitions are composed based on the output epsilons, and after that it is found that they are not required if the following output word label is not accepted by G. Specifically, when G accepts only a limited number of words by transitions from each state, many dead states and transitions can be composed, which remain in the composition result. But if G is made from an n-gram model, they are not so many, because an n-gram model usually allows any word to be connected each other, and therefore the n-gram WFST will accept any word in most states.

Next we consider another problem, which concerns the weight distribution along each path. As shown in Fig. 5.3, a non-zero weight that corresponds to an n-gram probability appears on a transition labeled with a word A, B, C, or D. A weight also appears on a back-off transition with

$\varepsilon : \varepsilon$. But most transitions have zero-valued weights, which are omitted from the figure. In the beam search for decoding, this weight distribution is ineffective because it cannot use those weights until a word label appears. This means that the WFST generated by on-the-fly composition does not have any look-ahead weights, which are useful for pruning unpromising hypotheses in an early stage of the search process. With fully static composition, L and G are first combined and then optimized. Since each word label in L is located at the first transition from the initial state as shown in Fig. 4.6, the word label and its weight are also located at the head of its phone sequence in $L \circ G$ (see Fig. 4.10). By determinizing $L \circ G$, the word labels and their weights are moved to the succeeding transitions, but the $\oplus$-sum of weights for bounded transitions is attached to each determinized transition (see Fig. 4.11), which plays the role of a look-ahead weight. Thus, the fully composed WFST N has look-ahead weights, which are effective for a beam search.

The above considerations suggest one possible way of improving the efficiency of decoding with on-the-fly composition, which provides a function that detects dead states and transitions and also obtains look-ahead weights simultaneously in an on-the-fly fashion. The next section introduces an approach to an advanced composition algorithm that includes this function.

5.4 LOOK-AHEAD COMPOSITION

Extension of the composition algorithm for on-the-fly use has been extensively investigated [CT01, CT06, CDD07, MSK07, ODIF09a, ARS09]. Those approaches aim to perform trimming and weight pushing on the fly in the composition operation simultaneously. This is basically accomplished by incorporating a *label look-ahead* mechanism in the original composition algorithm. The mechanism is usually implemented so that each state has a set of reachable non-epsilon labels from that state. We call such labels *prospective* labels, and let $\mathcal{L}_i(p)$ and $\mathcal{L}_o(p)$ be a set of prospective input labels and a set of prospective output labels of state p, respectively.

5.4.1 HOW TO OBTAIN PROSPECTIVE OUTPUT LABELS

Figure 5.4 shows WFST HCL, which is the same as the WFST in Fig. 5.1(a) but with a set of prospective output labels in each state. For example, label set {A,B,C,D} is assigned to state 1 because the paths from state 1 can reach transitions with an A, B, C, or D output label, i.e., $(2, \varepsilon, A, 0, 6)$, $(2, \varepsilon, C, 0, 6)$, $(3, \varepsilon, B, 0, 4)$, and $(3, \varepsilon, D, 0, 4)$. A meta symbol "<fin>" is introduced in some label sets, which denotes that the state is a final state or at least one path originating from the state can reach one final state.

These label sets are used to detect dead states and transitions and obtain look-ahead weights during the composition process. The label sets can be obtained efficiently using a depth-first search (DFS) algorithm, which starts from states that do not have any incoming transitions with output epsilon, and goes through transitions with epsilon output to those with non-epsilon output [CDD07].

This process, for example, can be accomplished by constructing a tree as in Fig. 5.5 and applying the DFS. We can construct the tree from the original WFST by making a unique leaf node

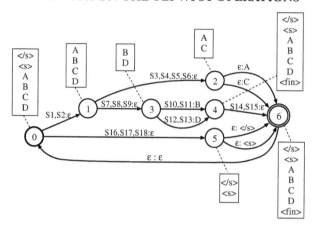

Figure 5.4: Example of WFST HCL with look-ahead labels: the set of look-ahead labels for each state is framed in a rectangle linked to the state with a dashed line.

for each transition with output label and redirecting the transition to the leaf node. Then, the states where we start the DFS are changed to root nodes. We may place a global root node for the multiple root nodes. The tree in Fig. 5.5 is constructed from the WFST in Fig. 5.4, where only state 4 is the root node. Once we construct the tree, the prospective label sets are efficiently obtained by collecting output labels and sending them to the parent nodes in the DFS process. In this way, we can make the prospective output labels for each state.

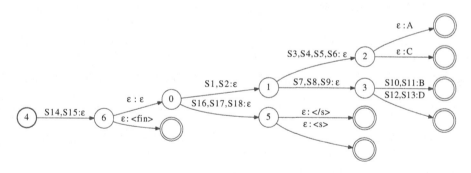

Figure 5.5: Tree structure constructed from the WFST in Fig. 5.4

5.4.2   BASIC PRINCIPLE OF LOOK-AHEAD COMPOSITION

Here, we explain the basic principle of the look-ahead composition. Suppose we perform the composition of two WFSTs $T_1 = (\Sigma_1, \Delta_1, Q_1, I_1, F_1, E_1, \lambda_1, \rho_1)$ and $T_2 =$

$(\Sigma_2, \Delta_2, Q_2, I_2, F_2, E_2, \lambda_2, \rho_2)$ using this method. In the algorithm, each state is composed of some two states, p_1 in Q_1 and p_2 in Q_2. The look-ahead composition allows us to combine states p_1 and p_2 only if the intersection of $\mathcal{L}_o(p_1)$ and $\mathcal{L}_i(p_2)$ is not empty. If the intersection is empty, this means that there will be no match between labels along the paths from p_1 and p_2. Consequently, the composed state (p_1, p_2) becomes a dead state. However, (p_1, p_2) is not a dead state if there are paths π_1 from p_1 and π_2 from p_2, which are both capable of reaching some final state with epsilons, i.e., $o[\pi_1] = \varepsilon, p[\pi_1] = p_1, n[\pi_1] \in F_1, i[\pi_2] = \varepsilon, p[\pi_2] = p_2$, and $n[\pi_2] \in F_2$. To avoid considering the composed state as a dead state, a special symbol indicating the capability of reaching some final state, e.g., <fin>, is also added to each prospective label set if the state is final or there is at least one path from the state to a final state.

In the look-ahead composition, we basically apply the following rules when composing a transition from an already composed state (p_1, p_2):

1. If $o[e_1] = i[e_2] \neq \varepsilon$ and $\mathcal{L}_o(n[e_1]) \cap \mathcal{L}_i(n[e_2]) \neq \emptyset$ for $e_1 \in E[p_1]$ and $e_2 \in E[p_2]$, compose transition $((p_1, p_2), i[e_1], o[e_2], w[e_1] \otimes w[e_2], (n[e_1], n[e_2]))$.

2. If $o[e_1] = \varepsilon$ and $\mathcal{L}_o(n[e_1]) \cap \mathcal{L}_i(p_2) \neq \emptyset$ for $e_1 \in E[p_1]$, compose transition $((p_1, p_2), i[e_1], \varepsilon, w[e_1], (n[e_1], p_2))$.

3. If $i[e_2] = \varepsilon$ and $\mathcal{L}_o(p_1) \cap \mathcal{L}_i(n[e_2]) \neq \emptyset$ for $e_2 \in E[p_2]$, compose transition $((p_1, p_2), \varepsilon, o[e_2], w[e_2], (p_1, n[e_2]))$.

Note that a composition filter is not assumed in the above rules, and therefore they may compose redundant transitions. We introduce a composition filter in the next section.

When we apply the look-ahead composition for HCL in Fig. 5.4 and G in Fig. 5.1(b), state $(1, 0)$ is not composed as appeared in Fig. 5.3 because the prospective label set {A,B,C} for state 1 in HCL does not include <s>, which is only the label acceptable from state 0 in G. States $(2, 0)$ and $(3, 0)$ are also successfully eliminated because state $(0, 1)$ is not composed. In contrast, state $(5, 0)$ is composed because the label set for state 5 in HCL includes <s>.

In a practical use of look-ahead composition in speech recognition, prospective input label sets $\mathcal{L}_i(p_2)$ for T_2 are not constructed. Instead $L_i(p_2)$ is used, which is a set of input labels of transitions outgoing from state p_2, i.e., $L_i(p_2) = \{i[e_2] | i[e_2] \neq \varepsilon, e_2 \in E[p_2]\}$. If $p_2 \in F_2$, we also add <fin> to $L_i(p_2)$, i.e., $L_i(p_2) = \{i[e_2] | i[e_2] \neq \varepsilon, e_2 \in E[p_2]\} \cup \{<fin>\}$. This simplification is for the following reason.

In WFST-based speech recognition with on-the-fly composition, WFSTs T_1 and T_2, which are to be combined, are usually HCL and G. Since word labels are moved to the later transitions in HCL as a result of determinization, label matching is delayed in the decoding process. Until label matching occurs, there is a potential for many dead states and transitions to be generated. Even if the composed states and transitions are coaccessible (not dead), they are searched without a help of the weights of G. This causes pruning errors in the Viterbi beam search. Thus, label look-ahead for HCL is important. On the other hand, if G is a WFST of an n-gram language model, most transitions in G do not have an input epsilon except for back-off transitions. Actually, each state has at most one epsilon transition for backing off, and such epsilon transitions continue at most $n - 1$ times, where n is usually 3 or 4. Accordingly, the operation does not generate many dead states and transitions caused by those input epsilons. Hence, the label look-ahead for G is not effective and the use of $\mathcal{L}_i(p_2)$ is avoided in speech recognition. This also helps to remove the overhead for obtaining $\mathcal{L}_i(p_2)$ when G is also composed on the fly, because states and transitions appearing beyond epsilons do not have to be composed just for looking ahead.

5.4.3   REALIZATION OF LOOK-AHEAD COMPOSITION USING A FILTER TRANSDUCER

As we mentioned above, look-ahead composition can be performed based on intersection $\mathcal{L}_o(p_1) \cap L_i(p_2)$, where $L_i(p_2)$ is used instead of $\mathcal{L}_i(p_2)$. However, if there are epsilon transitions from state p_2, $L_i(p_2)$ may not equal $\mathcal{L}_i(p_2)$ because $\mathcal{L}_i(p_2)$ potentially includes labels placed beyond the epsilon transitions while $L_i(p_2)$ includes only the immediate labels from state p_2. Accordingly, the decision in rule 2 may fail if there is at least one epsilon transition from state p_2. The condition of rule 2 can be relaxed so that if some $e_2 \in E[p_2]$ exists such that $i[e_2] = \varepsilon$, then the transition is composed independently of the result of $\mathcal{L}_o(n[e_1]) \cap L_i(p_2)$. But this relaxation may generate many dead states and transitions according to the structure of T_2. Oonishi et al. proposed an effective method that utilized the state indices of a composition filter [ODIF09a, ODIF09b]. By using this filter, epsilon input transitions in T_2 are not allowed immediately after an epsilon output transition in T_1. In this case, we can rely on the intersection result even if there exists epsilon transitions from state p_2.

In [ODIF09b], they used the two-state filter shown in Fig. 5.6. Note that this filter is slightly

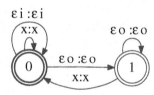

Figure 5.6: 2-state filter for look-ahead composition.

different from the filter presented in Fig. 3.11(a), where labels εi and εo are exchanged. The filter in Fig. 5.6 eliminates epsilon input transitions in T_2 immediately after an epsilon output transition in T_1, while that in Fig. 3.11(a) eliminates epsilon output transitions in T_1 immediately after an epsilon input transition in T_2. As we described above, the filter in Fig. 5.6 is more suitable for look-ahead composition.

When composing a transition from state (p_1, p_f, p_2) where each state is extended by p_f indicating the state of the filter FST, the rules for look-ahead composition are modified as follows.

1. If $o[e_1] = i[e_2] \neq \varepsilon$ for $e_1 \in E[p_1]$ and $e_2 \in E[p_2]$,
   compose transition $((p_1, p_f, p_2), i[e_1], o[e_2], w[e_1] \otimes w[e_2], (n[e_1], 0, n[e_2]))$.

2. If $o[e_1] = \varepsilon$ and $\mathcal{L}_o(n[e_1]) \cap L_i(p_2) \neq \emptyset$ for $e_1 \in E[p_1]$,
   compose transition $((p_1, p_f, p_2), i[e_1], \varepsilon, w[e_1], (n[e_1], 1, p_2))$.

3. If $p_f = 0$ and $i[e_2] = \varepsilon$ for $e_2 \in E[p_2]$,
   compose transition $((p_1, 0, p_2), \varepsilon, o[e_2], w[e_2], (p_1, 0, n[e_2]))$.

In the above rules, the intersection between the label sets is checked only in rule 2, because the filter FST makes a transition to state 1 where it obstructs epsilon input transitions in T_2, i.e., transition $e_2 \in E[p_2]$ such that $i[e_2] = \varepsilon$ is made only if $p_f = 0$. This corresponds to rule 3. However, the intersection is not checked in rule 1 since we can avoid obtaining label set $L_i(n[e_2])$ for destination state $n[e_2]$ to save the computation. To remove more dead states, we may also consider the intersection in rule 1 only if there is no epsilon input transitions from state $n[e_2]$.

5.4.4   LOOK-AHEAD COMPOSITION WITH WEIGHT PUSHING

We can also compose each transition with a weight pushed by the look-ahead mechanism. From some already composed state (p_1, p_2), transition $((p_1, p_2), i[e_1], \varepsilon, w[e_1], (n[e_1], p_2))$ is composed for $e_1 \in E[p_1]$ such that $o[e_1] = \varepsilon$. Such a transition has only weight $w[e_1]$ from T_1, and the composed path including such transitions does not have any weight from T_2 as long as the epsilon output continues. The look-ahead composition can incorporate such look-ahead weights as

$$w' = V[(p_1, p_2)]^{-1} \otimes V[(n[e_1], p_2)], \tag{5.2}$$

where $V[(s_1, s_2)]$ for a composed state (s_1, s_2) is the $\oplus$-sum of weights for future label matches, which can be obtained by referring to the prospective labels as

$$V[(s_1, s_2)] = \begin{cases} \left(\bigoplus_{e_2 \in E[s_2]:i[e_2] \in \mathcal{L}_o(s_1)} w[e_2] \right) \oplus \rho(s_2) & \text{if } \texttt{<fin>} \in \mathcal{L}_o(s_1) \text{ and } s_2 \in F_2 \\ \bigoplus_{e_2 \in E[s_2]:i[e_2] \in \mathcal{L}_o(s_1)} w[e_2] & \text{otherwise} \end{cases} \tag{5.3}$$

Here we assume that T_2 is epsilon free, i.e., $\forall e_2 \in E[s_2], i[e_2] \neq \varepsilon$. We finally obtain the composed transition as $((p_1, p_2), i[e_1], \varepsilon, w[e_1] \otimes w', (n[e_1], p_2))$. This is similar to the language model look-ahead technique [ONE96] that we mentioned in Section 2.7.3, which is widely used in traditional speech recognition approaches.

Look-ahead composition with weight pushing is also extended by the filter FST [ODIF09b] as follows.

1. If $o[e_1] = i[e_2] \neq \varepsilon$ for $e_1 \in E[p_1]$ and $e_2 \in E[p_2]$,
   compose transition $((p_1, p_f, p_2), i[e_1], o[e_2], w[e_1] \otimes w', (n[e_1], 0, n[e_2]))$, where $w' = V[(p_1, p_f, p_2)]^{-1} \otimes w[e_2]$.

2. If $o[e_1] = \varepsilon$ and $\mathcal{L}_o(n[e_1]) \cap L_i(p_2) \neq \emptyset$ for $e_1 \in E[p_1]$,
   compose transition $((p_1, p_f, p_2), i[e_1], \varepsilon, w[e_1] \otimes w', (n[e_1], 1, p_2))$, where $w' = V[(p_1, p_f, p_2)]^{-1} \otimes V[(n[e_1], 1, p_2)]$ and $V[(n[e_1], 1, p_2)]$ equals $V[(n[e_1], p_2)]$ obtained in Eq. (5.3).

3. If $p_f = 0$ and $i[e_2] = \varepsilon$ for $e_2 \in E[p_2]$,
   compose transition $((p_1, 0, p_2), \varepsilon, o[e_2], w[e_2], (p_1, 0, n[e_2]))$.

We assume that $V[(p_1, p_f, p_2)]$ is initialized by $\bar{1}$.

5.4.5 GENERALIZED COMPOSITION

Allauzen et al. proposed an efficient composition algorithm with the help of filters [ARS09], in which the filter function was generalized to include both the epsilon matching technique and the look-ahead mechanism. Algorithm 24 shows the pseudo-code, which is extended from the original pseudo-code in Algorithm 10 by considering a generic composition filter Φ. The filter Φ is not an FST as in Fig. 3.11 but it allows us to design a more powerful filter, for example, which blocks the composition of specific states and transitions and also modifies their labels and weights. Since Φ can be defined independently of the code, we can use an appropriate Φ depending on the purpose of composition.

The composition filter Φ is defined as

$$\Phi = (T_1, T_2, Q_f, i_f, \perp, \varphi, \rho_f), \tag{5.4}$$

where $Q_f, i_f, \perp$ and ρ_f denote a set of filter states, an initial filter state, a blocking filter state, and a final weight filter, respectively. φ is called a transition filter that blocks or modifies given transition pairs according to the current state of the filter.

In the original composition algorithm in Section 3.5, component WFSTs T_1 and T_2 are modified by adding self loops with εi or εo. Each output epsilon in T_1 and each input epsilon in T_2 are also replaced with εo and εi, respectively. These labels are introduced to handle epsilon matches explicitly in composition with a filter FST. Similarly, with the generalized composition, a self loop labeled with ε^L is added virtually to each state in T_1 and T_2, where ε^L is introduced for the composition filter. In fact, we prepare extended sets of transitions, $E^L[q_1] = E[q_1] \cup \{(q_1, \varepsilon, \varepsilon^L, \bar{1}, q_1)\}$ for all $q_1 \in Q_1$, and $E^L[q_2] = E[q_2] \cup \{(q_2, \varepsilon^L, \varepsilon, \bar{1}, q_2)\}$ for all $q_2 \in Q_2$.

In Algorithm 24, a composed state is represented as a triple (q_1, q_f, q_2), where $q_1 \in Q_1$ and $q_2 \in Q_2$ are states of T_1 and T_2, respectively. $q_f \in Q_f$ indicates the state of the filter.[2] In lines 1-4, a set of initial states I is created, where each composed state (i_1, i_f, i_2) is combined with the initial filter state i_f. In line 5, queue S and set of states Q are initialized by I. In lines 6-21, the expansion steps are performed for each composed state (q_1, q_f, q_2) in S. Line 9 checks whether the state is final or not. The final weight of the filter, $\rho_f(q_f)$, is also checked under this condition. This means

---

[2]In [ARS09] q_3 is used to represent a filter state, and (q_1, q_2, q_3) is used for a composed state. In this book, we use subscript f and state (q_1, q_f, q_2) to maintain consistency with the description for FST-based filters.

Algorithm 24 GeneralizedComposition(T_1, T_2, Φ)

1: **for** each $(i_1, i_f, i_2) \in I_1 \times \{i_f\} \times I_2$ **do**
2:     $\lambda((i_1, i_f, i_2)) \leftarrow \lambda_1(i_1) \otimes \lambda_2(i_2)$
3:     $I \leftarrow I \cup \{(i_1, i_f, i_2)\}$
4: **end for**
5: $Q \leftarrow S \leftarrow I$
6: **while** $S \neq \emptyset$ **do**
7:     $(q_1, q_f, q_2) \leftarrow \text{Head}(S)$
8:     $\text{Dequeue}(S)$
9:     **if** $(q_1, q_f, q_2) \in F_1 \times Q_f \times F_2$ and $\rho_f(q_f) \neq \bar{0}$ **then**
10:         $F \leftarrow F \cup \{(q_1, q_f, q_2)\}$
11:         $\rho((q_1, q_f, q_2)) \leftarrow \rho_1(q_1) \otimes \rho_f(q_f) \otimes \rho_2(q_2)$
12:     **end if**
13:     $M \leftarrow \{(e_1', e_2', q_f') = \varphi(e_1, e_2, q_f) | e_1 \in E^L[q_1], e_2 \in E^L[q_2], q_f' \neq \bot\}$
14:     **for** each $(e_1', e_2', q_f') \in M$ **do**
15:         **if** $(n[e_1'], q_f', n[e_2']) \notin Q$ **then**
16:             $Q \leftarrow Q \cup \{(n[e_1'], q_f', n[e_2'])\}$
17:             $\text{Enqueue}(S, (n[e_1'], q_f', n[e_2']))$
18:         **end if**
19:         $E \leftarrow E \uplus \{((q_1, q_f, q_2), i[e_1'], o[e_2'], w[e_1'] \otimes w[e_2'], (n[e_1'], q_f', n[e_2']))\}$
20:     **end for**
21: **end while**
22: **return** $T = (\Sigma_1, \Delta_2, Q, I, F, E, \lambda, \rho)$

---

that the filter can block the state from being a final. In line 13, M the set of triples (e_1', e_2', q_f') is obtained by using the transition filter φ, where blocking state $\bot$ is used to determine if each triple (e_1', e_2', q_f') should be included in M. If $q_f' = \bot$, the triple is excluded. In lines 14-20, a new transition is composed for each element in M.

Several composition filters have already been proposed in some studies undertaken by Allauzen et al. [ARS09, ARS11]. Composition with a filter FST and look-ahead composition with weight pushing can be realized using the following filters.

1. **Epsilon sequencing filter:** $\Phi_{\varepsilon-\text{seq}}$
   Let $Q_f = \{0, 1, \bot\}$, $i_f = 0$, $\rho_f(q_f) = \bar{1}$ for all $q_f \in Q_f$, and $\varphi(e_1, e_2, q_f) = (e_1, e_2, q_f')$ where:

$$q_f' = \begin{cases} 0 & \text{if } (o[e_1], i[e_2]) = (x, x) \text{ with } x \in \Delta_1 \equiv \Sigma_2 \\ 0 & \text{if } (o[e_1], i[e_2]) = (\varepsilon, \varepsilon^L) \text{ and } q_f = 0 \\ 1 & \text{if } (o[e_1], i[e_2]) = (\varepsilon^L, \varepsilon) \\ \bot & \text{otherwise} \end{cases} . \quad (5.5)$$

This is equivalent to the 2-state filter FST in Fig. 3.11(a), although different representations are used to handle epsilon transitions.

We can also define $\bar{\Phi}_{\varepsilon-\text{seq}}$ with

$$q'_f = \begin{cases} 0 & \text{if } (o[e_1], i[e_2]) = (x, x) \text{ with } x \in \Delta_1 \equiv \Sigma_2 \\ 0 & \text{if } (o[e_1], i[e_2]) = (\varepsilon^L, \varepsilon) \text{ and } q_f = 0 \\ 1 & \text{if } (o[e_1], i[e_2]) = (\varepsilon, \varepsilon^L) \\ \bot & \text{otherwise} \end{cases}, \tag{5.6}$$

which is symmetric with $\Phi_{\varepsilon-\text{seq}}$. This filter is equivalent to the filter FST in Fig. 5.6.

2. **Epsilon matching filter**: $\Phi_{\varepsilon-\text{match}}$
   Let $Q_f = \{0, 1, 2, \bot\}, i_f = 0, \rho_f(q_f) = \bar{1}$ for all $q_f \in Q_f$, and $\varphi(e_1, e_2, q_f) = (e_1, e_2, q'_f)$ where:

$$q'_f = \begin{cases} 0 & \text{if } (o[e_1], i[e_2]) = (x, x) \text{ with } x \in \Delta_1 \equiv \Sigma_2 \\ 0 & \text{if } (o[e_1], i[e_2]) = (\varepsilon, \varepsilon) \text{ and } q_f = 0 \\ 1 & \text{if } (o[e_1], i[e_2]) = (\varepsilon, \varepsilon^L) \text{ and } q_f \neq 2 \\ 2 & \text{if } (o[e_1], i[e_2]) = (\varepsilon^L, \varepsilon) \text{ and } q_f \neq 1 \\ \bot & \text{otherwise} \end{cases}. \tag{5.7}$$

This is equivalent to the 3-state filter FST in Fig. 3.11(b).

3. **Label reachability filter with weight pushing**: $\Phi_{\text{push-weight}}$
   Let $Q_f = \mathbb{K}$, $i_f = \bar{1}$, $\bot = \bar{0}$, $\rho_f(q_f) = q_f^{-1}$ for all $q_f \in Q_f$. and $\varphi(e_1, e_2, q_f) = (e_1, (p[e_2], i[e_2], o[e_2], w', n[e_2]), q'_f)$ where:

$$(q'_f, w') = \begin{cases} (\bar{1}, q_f^{-1}) & \text{if } o[e_1] = i[e_2] \\ (V[(n[e_1], q_2)], q_f^{-1} \otimes V[(n[e_1], q_2)] & \text{if } (o[e_1], i[e_2]) = (\varepsilon, \varepsilon^L) \\ (\bar{0}, w[e_2]) & \text{otherwise} \end{cases}. \tag{5.8}$$

Note that each filter state is identified by a $\oplus$-sum of weights and is used to obtain the transition weight w'. This filter can be used for look-ahead composition with weight pushing.

The benefit of composition filters is that a new filter can be synthesized by combining existing filters. The literature [ARS11] shows how to generate a new filter from two composition filters.

Suppose $\Phi^a = (Q_f^a, i_f^a, \bot^a, \varphi^a, \rho_f^a)$ and $\Phi^b = (Q_f^b, i_f^b, \bot^b, \varphi^b, \rho_f^b)$ are two composition filters. The combined filter $\Phi^a \diamond \Phi^b = (Q_f, i_f, \bot, \varphi, \rho_f)$ is defined as

$$Q_f = Q_f^a \times Q_f^b, i_f = (i_f^a, i_f^b), \bot = (\bot^a, \bot^b), \rho((q_f^a, q_f^b)) = \rho_f^a(q_f^a) \otimes \rho_f^b(q_f^b),$$

and for a given (e_1, e_2, q_f)

$$\varphi(e_1, e_2, q_f) = (e_1'', e_2'', q_f') \text{ with } q_f' = \begin{cases} \bot & \text{if } r_f^a = \bot^a \text{ or } r_f^b = \bot^b \\ (r_f^a, r_f^b) & \text{otherwise} \end{cases},$$

where $q_f = (q_f^a, q_f^b)$, $\varphi(e_1, e_2, q_f^b) = (e_1', e_2', r_f^b)$, and $\varphi(e_1', e_2', q_f^a) = (e_1'', e_2'', r_f^a)$. For example, a combined filter $\Phi_{\text{push-weight}} \diamond \bar{\Phi}_{\varepsilon-\text{seq}}$ will perform the look-ahead composition in Section 5.4.4.

5.4.6   INTERVAL REPRESENTATION OF LABEL SETS

Although the look-ahead composition technique effectively eliminates dead states/transitions and obtains pushed weights during decoding, the computational overhead for this look-ahead processing may become very large if we do not implement it carefully. For example, the size of a prospective label set $|\mathcal{L}_o(p)|$ for some $p \in Q_1$ can be very large in large-vocabulary speech recognition. The maximum size can equal the vocabulary size if T_1 is HCL. Therefore, the intersection $\mathcal{L}_o(p_1) \cap L_i(p_2)$ for some $p_1 \in Q_1$ and $p_2 \in Q_2$ may need a large amount of computation. In addition, summing up the weights to obtain a pushed weight as in Eq. (5.3) may also take a long time because the enumeration of all $e_2 \in E[p_2]$ such that $i[e_2] \in \mathcal{L}_o(n[e_1])$ can be expensive for large $\mathcal{L}_o(n[e_1])$ and large $L_i(p_2)$. Therefore, efficient implementation is important as regards minimizing the overhead.

Allauzen et al. proposed using an interval representation of label sets, which was also useful for computing the pushed weights efficiently [ARS09]. The interval representation is used to define a label set as a pair of label Id numbers, i.e., $\mathcal{L}_o(p_1) \equiv [b_o(p_1), e_o(p_1))$ for all $p_1 \in Q_1$, where $b_o(p_1)$ and $e_o(p_1)$ denote label Id numbers indicating the starting Id and the ending Id+1 in $\mathcal{L}_o(p_1)$, respectively, i.e., $\forall l \in \mathcal{L}_o(p_1)$, $b_o(p_1) \le l < e_o(p_1)$, and $\forall r \notin \mathcal{L}_o(p_1)$, $r < b_o(p_1)$ or $e_o(p_1) \le r$. To use the interval representation, we have to renumber all the output labels in T_1 so that every prospective label set for each $p_1 \in Q_1$ is represented in the form of an interval. Then, the input labels in T_2 are also renumbered by the new Id numbers. This renumbering can be done in advance of decoding.

In the look-ahead composition process, whether $\mathcal{L}_o(p_1) \cap L_i(p_2)$ is empty or not is efficiently evaluated by checking whether there is at least one element $l \in L_i(p_2)$ such that $b_o(p_1) \le l < e_o(p_1)$. This check can be performed by computing the upper and lower bounds of the interval in $L_i(p_2)$, whose complexity is $O(\log_2 |L_i(p_2)|)$ as determined using the binary search, where all the transitions for each state in T_2 should already be sorted by the new Id numbers.

In addition, we can obtain the $\oplus$-sum of the weights as

$$
\begin{aligned}
V[(p_1, p_2)] &= \bigoplus_{e_2 \in E[p_2]: i[e_2] \in \mathcal{L}_o(p_1)} w[e_2] \\
&= W(p_2, e_o(p_1)) \oplus_{\text{diff}} W(p_2, b_o(p_1)),
\end{aligned}
\tag{5.9}
$$

where $\oplus_{\text{diff}}$ computes the $\oplus$-difference between two arguments. $W(s, n)$ is the cumulative weight for all $w[e]$ such that $e \in E[s]$ and $i[e] < n$. $W(s, n)$ can be pre-computed for each (s, n) pair as

$$
W(s, n) = \bigoplus_{e \in E[s]: i[e] < n} w[e],
\tag{5.10}
$$

where n is limited such that $n \in L_i(s) \cup \{(\max_{e \in E[s]} i[e]) + 1\}$. Thus, the computation for $V[(p_1, p_2)]$ in Eq. (5.9) has a very short run time.

A final problem remains, which is how to renumber all the output labels in the first WFST to introduce the interval representation. Allauzen et al. stated that the problem could be considered a consecutive ones problem (C1P). Given a set Σ in which each element belongs to some subset $\mathcal{L} \subseteq \Sigma$, C1P is solved by ordering all the elements in Σ such that the elements are also consecutive in each subset $\mathcal{L}$ with respect to the order.

Suppose that the output labels of WFST HCL in Fig. 5.4 are originally numbered as in Table 5.1, where we also add a final state symbol <fin>. The prospective label sets over all the states

Table 5.1: Original label Ids for HCL

label	Id
</s>	1
<s>	2
A	3
B	4
C	5
D	6
<fin>	7

in HCL can be represented as a 0/1 matrix as shown in Fig. 5.7, where each row corresponds to a state and each column corresponds to a label. Each 1 or 0 value in the matrix indicates whether or not

		label					
state		</s>	<s>	A	B	C	D	<fin>
0	1	1	1	1	1	1	0
1	0	0	1	1	1	1	0
2	0	0	1	0	1	0	0
3	0	0	0	1	0	1	0
4	1	1	1	1	1	1	1
5	1	1	0	0	0	0	0
6	1	1	1	1	1	1	1

Figure 5.7: 0/1 matrix representing prospective label sets

the label is a member of the prospective label set for the state. Accordingly, to represent each label set as an interval, a block of consecutive ones must appear only once in each row. This is called the *consecutive ones property*. The C1P is the problem of obtaining all the rows that have such consecutive ones by permutating the columns of the matrix. Since the matrix in Fig. 5.7 has some gaps between the ones in the rows for states 2 and 3, it does not satisfy the property.

Although there are some general algorithms for solving the C1P such as the PQ-tree algorithm [BL76], the solution can also be derived from a depth-first search (DFS) for WFSTs used in speech recognition. If we apply the DFS to the WFST in Fig. 5.4 in the same way as that for obtaining the prospective label sets, the labels can be ordered and numbered as in Table 5.2. This order

Table 5.2: New label Ids for HCL	
label	Id
A	1
C	2
B	3
D	4
</s>	5
<s>	6
<fin>	7

is actually the same as the depth-first order of output labels in the tree of Fig. 5.5. By permutating the columns in the matrix according to the obtained order, the consecutive ones property is satisfied in Fig. 5.8. As a result, each prospective label set for each state can be represented as the intervals in Table 5.3.

		label						
		A	C	B	D	</s>	<s>	<fin>
state	0	1	1	1	1	1	1	0
	1	1	1	1	1	0	0	0
	2	1	1	0	0	0	0	0
	3	0	0	1	1	0	0	0
	4	1	1	1	1	1	1	1
	5	0	0	0	0	1	1	0
	6	1	1	1	1	1	1	1

Figure 5.8: A solution for the C1P

However, C1P cannot be solved for all cases, i.e., there are matrices for which a permutation yielding the consecutive ones property is not found. The C1P for WFST L can be solved if a unique pronunciation is assigned to each word label. To satisfy this condition, when a word has multiple pronunciations, the word label has to be extended by subscripting those pronunciations. The permutation can also be obtained for the determinized and minimized WFST of L. But, even if the C1P for L can be solved, the C1P for the WFST composed of C and L may not be solved. In

Table 5.3: Interval representation of prospective label sets for HCL

state	prospective label set	interval
0	{A,C,B,D, \</s>, \<s>}	$[1, 7)$
1	{A,C,B,D}	$[1, 5)$
2	{A,C}	$[1, 3)$
3	{B,D}	$[3, 5)$
4	{A,C,B,D, \</s>, \<s>, \<fin>}	$[1, 8)$
5	{\</s>, \<s>}	$[5, 7)$
6	{A,C,B,D, \</s>, \<s>, \<fin>}	$[1, 8)$

such a case, intervals obtained for L may be used for CL and HCL instead, where the interval for each state in L is assigned to each composed state associated with the original state number in L.

5.5 ON-THE-FLY RESCORING APPROACH

Another approach to fast and memory-efficient decoding in WFST-based speech recognition is on-the-fly rescoring [HHM04, HN05, HHMN07]. The look-ahead composition approach focuses on how to dynamically generate states and transitions of a fully composed and optimized WFST. Since a dynamically composed WFST can be considered a single WFST, the decoding algorithm does not need to be changed depending on whether the WFST is generated statically or dynamically. On the other hand, the on-the-fly rescoring approach includes a composition step in the decoding process, but actually does not make the fully composed WFST. With this method, the first component WFST is simply used to generate hypothetical output label sequences from the WFST, which are then combined with the next WFSTs to rescore the hypotheses. Thus, this method is designed to compose hypotheses but not WFSTs. In the following sections, we describe the concept and its algorithm.

5.5.1 CONSTRUCTION OF COMPONENT WFSTS FOR ON-THE-FLY RESCORING

The on-the-fly rescoring approach employs an implementation based on incremental language models (incremental LMs) [DH01, WMMK01]. For example, a set of WFSTs for speech recognition, $\{H, C, L, G\}$, is used to construct two WFSTs, $HCLG_m$ and $G_{n/m}$, where H, C, L, and G are WFSTs of HMMs, triphone context dependency, a pronunciation lexicon, and an n-gram language model, respectively. But G is decomposed into G_m and $G_{n/m}$ such that $G = G_m \circ G_{n/m}$ for some m where $1 \le m < n$. G_m is the WFST of the m-gram language model. $G_{n/m}$ is the WFST of the n-gram language model where each n-gram probability is divided by the corresponding m-gram probability as follows

$$P'(w_n|w_1^{n-1}) = \frac{P(w_n|w_1^{n-1})}{P(w_n|w_{n-m+1}^{n-1})}. \qquad (5.11)$$

The transition weights in $G_{n/m}$ are given as the minus logarithm of the divided probabilities.

WFST $HCLG_m$ can be composed and optimized in the same way as WFST N in Eq. (4.9). $HCLG_m$ and $G_{n/m}$ are combined on the fly, where $HCLG_m \circ G_{n/m} = N$. Thus, the incremental LM approach decomposes a language model into multiple WFSTs with increasing complexities and combines them by on-the-fly composition. The purpose of the incremental LM is to apply the language model scores as soon as possible before label matches are realized by the composition operation. The motivation is similar to that for the look-ahead composition with weight pushing, which can improve the pruning efficiency in a beam search. However, the statically pushed weights in $HCLG_m$ are limited to the lower-order probabilities given by the m-gram model. If we assume $n = 3$ and $m = 1$ as in [WMMK01], only unigram look-ahead weights are available, and those unigram weights are replaced with trigram weights when each path reaches a transition with a word label. Since the unigram probabilities are less accurate than the trigram probabilities as a language model, the beam search is less effective than that with look-ahead composition of the trigram WFST. Although the on-the-fly rescoring approach does not use dynamic weight pushing, it reduces the decoding time in a different way.

In standard (or look-ahead) on-the-fly composition, two or more WFSTs are combined during decoding. The Viterbi search is performed for the composed WFST. With the rescoring approach, the Viterbi search is performed based on the first WFST. The second or higher WFSTs are only used to rescore the paths being searched (i.e., hypotheses) in the first WFST. Since this rescoring is performed efficiently in an on-the-fly fashion, the total amount of computation is almost the same as when using only the first WFST. Thus, the algorithm can achieve a faster and more memory-efficient search than that with the standard on-the-fly composition. Note that the search process is completely one-pass, i.e., the rescoring step is performed frame by frame as necessary in the one-pass procedure. In addition, the algorithm achieves not only rescoring but also transduction with multiple WFSTs in a one-pass manner.

5.5.2 CONCEPT

We first define terms for describing the concept of the on-the-fly rescoring approach. We define a path along consecutive transitions from the initial state as a hypothesis $h = e_1, \ldots, e_k$, where $p[e_1] = i$ for some $i \in I$, $n[e_{j-1}] = p[e_j]$, $j = 2, \ldots, k$. We extend $n[\cdot]$ and $p[\cdot]$ to paths as $n[h] = n[e_k]$ and $p[h] = p[e_1]$. We also extend $o[\cdot]$ and $w[\cdot]$ to paths as $o[h] = o[e_1] \cdots o[e_k]$ and $w[h] = w[e_1] \otimes \cdots \otimes w[e_k]$. If $n[e_k] \in F$ after an input symbol sequence X was accepted (i.e. $i[h] = X$), h is a successful hypothesis. Otherwise, we call h a partial hypothesis or just a hypothesis. In addition, given a WFST T and symbol sequences X and Y, we use $\Pi_T(X, Y)$ to denote a set of successful paths that accept X and output Y with T. We also use $\Pi_T(X) = \bigcup_Y \Pi_T(X, Y)$. The overall weight of T for an input-output sequence (X, Y) is computed as

$$[[T]](X, Y) = \bigoplus_{\pi \in \Pi_T(X,Y)} \lambda(p[\pi]) \otimes w[\pi] \otimes \rho(n[\pi]). \tag{5.12}$$

Suppose the decoder employs two WFSTs $T_1 = (\Sigma, \Delta, Q_1, I_1, F_1, E_1, \lambda_1, \rho_1)$ and $T_2 = (\Delta, \Gamma, Q_2, I_2, F_2, E_2, \lambda_2, \rho_2)$, which are combined on the fly. In standard on-the-fly composition approaches, given an input symbol sequence X, the decoder finds $\hat{Z}$ such that

$$[[T_1 \circ T_2]](X, \hat{Z}) = \min_{Z \in \Gamma^*} [[T_1 \circ T_2]](X, Z) \tag{5.13}$$

$$= \min_{(Y,Z) \in \Delta^* \times \Gamma^*} [[T_1]](X, Y) \otimes [[T_2]](Y, Z), \tag{5.14}$$

where we assume the use of a tropical semiring, i.e., the decoder searches for the minimally weighted path that outputs $\hat{Z}$. We can rewrite Eqs. (5.13) and (5.14) as those for finding the best path as

$$[[T_1 \circ T_2]](X, \hat{Z}) = \min_{\pi \in \Pi_{T_1 \circ T_2}(X,Z)} \lambda_{1 \circ 2}(p[\pi]) \otimes w[\pi] \otimes \rho_{1 \circ 2}(n[\pi]) \tag{5.15}$$

$$= \min_{(h,f) \in \Pi_{T_1}(X,Y) \times \Pi_{T_2}(Y):Y \in \Delta^*} (\lambda_1(p[h]) \otimes \lambda_2(p[f])) \otimes (w[h] \otimes w[f])$$

$$\otimes (\rho_1(n[h]) \otimes \rho_2(n[f])), \tag{5.16}$$

where $\lambda_{1 \circ 2}(\cdot)$ and $\rho_{1 \circ 2}(\cdot)$ denote the initial and final weight functions of $T_1 \circ T_2$, respectively. According to the above equations, finding the best path in $T_1 \circ T_2$ is equivalent to finding the best pair of paths h in T_1 and f in T_2 such that $o[h] = i[f]$ whose label sequence corresponds to Y.

We note here that (5.14) can be rewritten as

$$[[T_1 \circ T_2]](X, \hat{Z}) = \min_{Y \in \Delta^*} \left\{ [[T_1]](X, Y) \otimes \min_{Z \in \Gamma^*} [[T_2]](Y, Z) \right\}. \tag{5.17}$$

This equation implies that the algorithm for finding $\hat{Z}$ given X could be reorganized based on an algorithm for finding Y that derives $\min_{Y \in \Delta^*}[[T_1]](X, Y)$ with a slight modification where $[[T_1]](X, Y)$ is adjusted by the $\otimes$-product of $\min_{Z \in \Gamma^*}[[T_2]](Y, Z)$. We can also rewrite Eq. 5.17 with the paths as

$$[[T_1 \circ T_2]](X, \hat{Z}) = \min_{h \in \Pi_{T_1}(X)} \lambda_1(p[h]) \otimes w[h] \otimes \rho_1(n[h])$$

$$\otimes \left\{ \min_{f \in \Pi_{T_2}(o[h])} \lambda_2(p[f]) \otimes w[f] \otimes \rho_2(n[f]) \right\}. \tag{5.18}$$

Since the decoder finds the minimally weighted path in T_2 only for each $Y (= o[h])$, it is not necessary to consider all possible combinations of paths as in Eq. (5.16). This fact reveals the potential for reducing the amount of computation needed for a search with the composed WFST.

According to Eq. (5.18), the search procedure may consist of two steps: (1) enumeration of all $Y (= o[h])$ on T_1's paths accepting X, and (2) selection of the minimally weighted path in T_2's paths accepting Y. In practice, this procedure can be performed with a two-pass strategy, i.e., lattice generation by T_1 and rescoring of the lattice by T_2, where a WFST is generated by composition of the lattice and T_2, and a shortest path search is applied to the WFST. Since the first pass (lattice generation) uses only T_1, the amount of computation it requires is much less than that of $T_1 \circ T_2$. Furthermore, if pruning techniques are used in the first pass, the size of the lattice can be reduced and the amount of computation needed for the second pass can also be reduced. The idea of the on-the-fly rescoring approach is to solve Eq. (5.18) with a one-pass search strategy.

In the algorithm, partial hypotheses are generated from T_1 but they are weighted based on Eq. (5.18) during a time-synchronous Viterbi search. The main difference from two-pass search strategies is that the decoder uses all knowledge sources, i.e., both WFSTs T_1 and T_2, from the beginning of the search. This is effective for both correct path selection and incorrect path pruning. Compared with the standard approach that handles the search network based on $T_1 \circ T_2$, the rescoring approach utilizes a smaller network based only on T_1. Hence, the rescoring-based method essentially requires less computation than the standard method.

Suppose that there is a partial hypothesis h generated from T_1 and that its weight is computed as $\lambda_1(p[h]) \otimes w[h]$. In the rescoring approach, h is linked to a set of co-hypotheses generated from T_2 by taking the label sequence $o[h]$ as T_2's input, where we call the hypotheses produced by T_2 "co-hypotheses" to distinguish them from the hypotheses produced by T_1. The decoder rescores h using the minimally weighted co-hypothesis in the set. The rescoring is performed by handling a list of co-hypotheses $g[h]$ associated with each hypothesis h. During the decoding, the following basic procedure is used to generate each hypothesis and rescore it with its co-hypotheses.

When a new hypothesis h' is generated by adding a transition e originating from state $n[h]$, the cumulative weight of h' is basically obtained as $\lambda_1(p[h]) \otimes w[h] \otimes w[e]$. If the transition e outputs nothing ($o[e] = \varepsilon$ and $o[h'] = o[h]$), no new co-hypothesis is generated with T_2. In this case, the list of co-hypotheses remains as it is, i.e., $g[h'] = g[h]$. Only when the transition e outputs a non-epsilon symbol y ($o[e] = y \neq \varepsilon$ and $o[h'] \neq o[h]$), a new co-hypothesis f' is generated for each existing co-hypothesis f in $g[h]$ by adding a transition r, which originates from state $n[f]$ with an input symbol y. The weight of f' is calculated as $\lambda_2(p[f]) \otimes w[f] \otimes w[r]$ as well. The new co-hypotheses are then stored in $g[h']$. We then use the following cumulative weight for hypothesis h' as

$$\alpha_1(h') = \lambda_1(p[h']) \otimes w[h'] \otimes \min_{f' \in g[h']} \lambda_2(p[f']) \otimes w[f']. \tag{5.19}$$

For simplicity, we introduce a joint cumulative weight

$$\alpha_2(h', f') = (\lambda_1(p[h']) \otimes w[h']) \otimes (\lambda_2(p[f']) \otimes w[f'], \tag{5.20}$$

for co-hypothesis $f' \in g[h']$, and Eq. (5.19) can simply be rewritten as

$$\alpha_1(h') = \min_{f' \in g[h']} \alpha_2(h', f'). \tag{5.21}$$

After processing the final symbol of the input sequence, the best successful hypothesis and the best successful co-hypothesis can be simultaneously derived as

$$(\hat{h}, \hat{f}) = \underset{h, f : f \in g[h], n[h] \in F_1, n[f] \in F_2}{\operatorname{argmin}} \alpha_2(h, f) \otimes \rho_1(n[h]) \otimes \rho_2(n[f]) \tag{5.22}$$

Eventually, the result of the search for the best symbol sequence is given by $o[\hat{f}]$, which is the output symbol sequence of the best successful co-hypothesis.

Figure 5.12 shows the decoding process in the rescoring approach when using WFSTs $HCLG_m$ and $G_{n/m}$ from Figs. 5.9 and 5.10 as an example. The upper half of Fig. 5.12 repre-

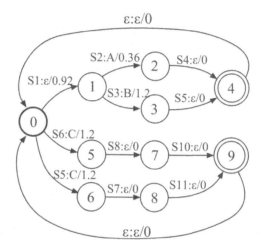

Figure 5.9: An example of WFST $HCLG_m$: input labels S1,S2, ... ,S11 are HMM-state Ids, and output labels A, B, C represent distinct words.

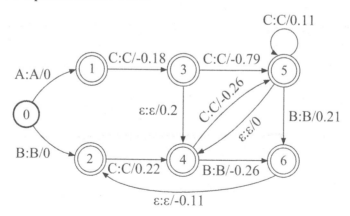

Figure 5.10: An example of WFST $G_{n/m}$: a, b, and c represent distinct words.

sents a set of partial hypotheses generated with $HCLG_m$. The lower half of Fig. 5.12 shows that each hypothesis is linked to a list of co-hypotheses generated with $G_{n/m}$, and is rescored using the co-hypotheses in the list. Since minimal computation is required to update the list of co-hypotheses, the total amount of computation is almost the same as when decoding with only the first WFST. On the other hand, Fig. 5.11 represents hypotheses generated with the composite WFST of $HCLG_m$ and $G_{n/m}$. Here the number of hypotheses dealt with at each time frame is much larger than that for the rescoring approach in Fig. 5.12. Thus, the basic Viterbi search is performed based only on

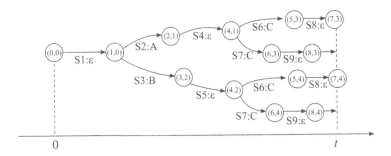

Figure 5.11: Hypotheses generated during decoding with a standard composition algorithm: each path from the initial state (0,0) represents a hypothesis at the current time frame t.

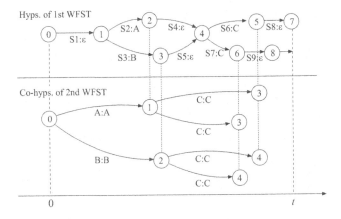

Figure 5.12: Hypotheses and co-hypotheses generated during decoding with the on-the-fly rescoring approach

the first WFST in the rescoring approach. Additional computation for updating the co-hypothesis lists is only necessary when a hypothesis is expanded by a transition with a non-epsilon output label. If the first and second WFSTs are designed as $HCLG_m$ and $G_{n/m}$, most transitions in the first WFST have an epsilon output label. Accordingly, little computation is needed for rescoring each hypothesis.

Algorithm   25   ViterbiBeamSearchWithOnTheFlyRescoring(N = {$N_m | m = 1, \ldots, M$}, X = $x[1], \ldots, x[T]$)

1:   $S \leftarrow$ initialize($\mathbf{I}, \Lambda$)
2:   $A \leftarrow \emptyset$
3:   **for** $t = 1$ to T **do**
4:       $S \leftarrow$ transition_with_epsilon($\mathbf{E}, S, t - 1$)
5:       $\langle S, A \rangle \leftarrow$ transition_with_input($\mathbf{E}, S, A, X, t$)
6:       prune(S, A, t)
7:   **end for**
8:   $\hat{B} \leftarrow$ final_transition($\mathbf{E}, \mathbf{F}, \mathbf{R}, S, T$)
9:   $Y =$ backtrack($\hat{B}$)
10: **return** Y

Note that, when the Viterbi search chooses the best partial hypothesis from a set of active hypotheses that meet at a state in T_1, their co-hypothesis lists must be merged into one list. If different co-hypotheses have reached the same state in T_2, only the best co-hypothesis among them is retained in the merged list. Thus, a certain overhead is inevitable due to the list processing, but it is nonsignificant in the total computation of the decoding. Moreover, lists do not need to be merged when the two co-hypothesis lists are identical, and the frequency of the merging is reduced by using the approximation described in Section 5.5.4.

5.5.3   ALGORITHM

The algorithm of on-the-fly rescoring can be expressed as an extension of the decoding algorithm using a single WFST found in Section 4.4 (Algorithms 17–23). The basic procedure is the same as the original when using only the first WFST of multiple WFSTs but a few additional steps are inserted for rescoring hypotheses with the remaining WFSTs. Although we explained the concept of the rescoring approach for two WFSTs, the algorithm presented here covers cases where these are more than two WFSTs.

Let M be the number of WFSTs used in the algorithm, and N_m be the m-th WFST where $N_m = (\Sigma_m, \Delta_m, Q_m, I_m, F_m, E_m, \lambda_m, \rho_m)$ and $1 \leq m \leq M$. We also use the notations $\mathbf{N} = \{N_m | m = 1, \ldots, M\}$, $\mathbf{I} = \{I_m | m = 1, \ldots, M\}$, $\mathbf{F} = \{F_m | m = 1, \ldots, M\}$, $\mathbf{E} = \{E_m | m = 1, \ldots, M\}$, $\Lambda = \{\lambda_m | m = 1, \ldots, M\}$, and $\mathbf{R} = \{\rho_m | m = 1, \ldots, M\}$ to represent the components of the M WFSTs. Algorithm 25 shows a pseudo-code of the main part of the algorithm. This is actually the same as Algorithm 17 except that it uses the sets $\mathbf{N}, \mathbf{I}, \mathbf{F}, \mathbf{E}, \Lambda$, and $\mathbf{R}$ instead of N, I, F, E, λ, and ρ for a single WFST.

In the sub-codes called from the main code, we use the following quantities.

$\alpha_m(t, s)$: Cumulative weight of a partial path up to frame t at a WFST state s, which corresponds to $\alpha_1(h')$ in Eq. (5.19) and $\alpha_2(h', f')$ in Eq. (5.20) for hypothesis h' and co-hypothesis f'. Note

Algorithm 26 initialize($\mathbf{I}, \Lambda$)

1: **for** each $i \in I_1$ **do**
2:     $\langle \lambda', g' \rangle \leftarrow$ initial_rescore($\mathbf{I}, \Lambda, \lambda_1(i), i, 2$)
3:     $\alpha_1(0, i) \leftarrow \lambda'$
4:     $g_1(0, i) \leftarrow g'$
5:     $B_1(0, i) \leftarrow \langle 0, 0 \rangle$
6:     Enqueue(S, i)
7:     **return** S
8: **end for**

---

that state s represents a combined state consisting of the m states that the 1-to-m WFSTs have reached.

$B_m(t, s)$: Back pointer to keep track of the most likely state sequence up to frame t at a WFST state s. $B(t, s)$ takes a pair $\langle \tau, e \rangle$, where τ indicates the starting frame of the transition e, where e is the most likely transition incoming to state s at frame t. Let $e = 0$ if there is no transition incoming to state s.

$g_m(t, s)$: Co-hypothesis list associated with a (co)hypothesis up to frame t at a WFST state s. $g_m(t, s)$ takes a pair $\langle \tau, q \rangle$ as a component of the list, which means that the co-hypothesis for m-the WFST has arrived in state q at frame τ, where q represents a combined state.

$\alpha(t, e, j)$: Cumulative weight of a partial path up to frame t at an HMM state j in transition e.

$b(t, e, j)$: Back pointer to keep track of the most likely state sequence up to frame t at an HMM state j in transition e. $b(t, e, j)$ holds only the starting frame of the transition e.

The pseudo-code, initialize($\mathbf{I}, \Lambda$), is shown in Algorithm 26. A set of initial states, I_1 and the initial weight function λ_1 of the first WFST N_1 are used to generate initial hypotheses at time frame 0. On line 2, a function initial_rescore($\mathbf{I}, \Lambda, \lambda, i, m$) is called to obtain a rescored weight λ' and a co-hypothesis list g' for each initial hypothesis. Then, λ' and g' are assigned to the cumulative weight of the hypothesis, $\alpha_1(i.i)$ and its co-hypothesis list $g_1(0, i)$ in lines 3 and 4, respectively. The back pointer $B_1(0, i)$ is initialized on line 5. Then each initial state is inserted in queue S on line 6 to extend the hypotheses from the state in the following steps.

The pseudo-code of initial_rescore() is shown in Algorithm 27. This function takes the $\otimes$-product of initial weights, λ and the composed state i up to the $(m-1)$-th WFSTs in addition to $\mathbf{I}$, Λ and m, and returns the rescored weight $\hat{\lambda}$ and the co-hypothesis list $\hat{g}$ for the $(m-1)$-th WFST. As in lines 4–9, the function is called recursively until $m = M$. The cumulative weight and back pointer of each co-hypothesis are initialized in lines 10 and 11. In lines 12–14, the best initial weight among those of the co-hypotheses is obtained as $\hat{\lambda}$. The co-hypothesis list $\hat{g}$ is constructed on line 15.

Algorithm 27 initial_rescore($\mathbf{I}, \Lambda, \lambda, i, m$)

1: $\hat{\lambda} \leftarrow \bar{0}$
2: $\hat{g} \leftarrow \emptyset$
3: **for** each $i' \in I_m$ **do**
4:     **if** $m < M$ **then**
5:         $\langle \lambda', g' \rangle \leftarrow$ initial_rescore$(\mathbf{I}, \Lambda, \lambda \otimes \lambda_m(i'), (i, i'), m + 1)$
6:         $g_m(0, (i, i')) \leftarrow g'$
7:     **else**
8:         $\lambda' \leftarrow \lambda \otimes \lambda_m(i')$
9:     **end if**
10:     $\alpha_m(0, (i, i')) \leftarrow \lambda'$
11:     $B_m(0, (i, i')) \leftarrow \langle 0, 0 \rangle$
12:     **if** $\hat{\lambda} \oplus \lambda' = \lambda'$ **then**
13:         $\hat{\lambda} \leftarrow \lambda'$
14:     **end if**
15:     Enqueue$(\hat{g}, \langle 0, (i, i') \rangle)$
16: **end for**
17: **return** $\langle \hat{\lambda}, \hat{g} \rangle$

The pseudo-code of transition_with_epsilon() for the rescoring approach is shown in Algorithm 28, where epsilon transitions are made from each state in queue S, and new hypotheses for the destination states are made with their weights and back pointers. This pseudo-code is similar to Algorithm 19 for decoding with a single WFST except that function rescore() is called on line 7 and assignments of rescored weight α' to $\alpha_1(t, n[e])$ and co-hypothesis list g' to $g_1(t, n[e])$ in the following lines. Note that $E_1, \alpha_1(t, n[e])$ and $B_1(t, n[e])$ have subscript "1" in the pseudo-code, which indicates the first of the M WFSTs. Hence, the computation amount of this algorithm is almost the same as that for decoding with only the first WFST if we ignore the computation for rescore().

We show the pseudo-code of transition_with_input() in Algorithm 29, which is also similar to Algorithm 20 for decoding with a single WFST. In lines 1–10, the hypotheses go into the HMM states of each transition outgoing from each state s in S. The difference from the original is only line 5 where the co-hypothesis list $g_1(t - 1, s)$ is taken over by $g(t - 1, e, i_e)$. In lines 11–26, HMM-level state transitions are made for each pair $\langle e, j \rangle$ in A, where the only difference is line 20 for taking over the co-hypothesis list. In lines 27–38, the cumulative weight and the back pointer outgoing from each transition e at frame t are given to the next WFST state, where function rescore() is called and rescored weight α' is assigned to $\alpha_1(t, n[e])$ and co-hypothesis list g' is assigned to $g_1(t, n[e])$.

The function rescore() is shown in Algorithm 30, which is a major component of the rescoring approach. This function takes arguments $\mathbf{E}, \alpha, g, e, t$, and m, where α is the $\otimes$-product of cumulative

---

Algorithm 28 transition_with_epsilon($\mathbf{E}, S, t$)

1: $S' \leftarrow \emptyset$
2: **while** $S \neq \emptyset$ **do**
3:      $s \leftarrow \text{Head}(S)$
4:      Dequeue(S)
5:      **for each** $e \in E_1(s, \varepsilon)$ **do**
6:          $\alpha' \leftarrow \alpha_1(t, s) \otimes w[e]$
7:          $\langle \alpha', g' \rangle \leftarrow \text{rescore}(\mathbf{E}, \alpha', g_1(t, s), e, t, 2)$
8:          **if** $\alpha_1(t, n[e]) \oplus \alpha' = \alpha'$ **then**
9:              $\alpha_1(t, n[e]) \leftarrow \alpha'$
10:             $B_1(t, n[e]) \leftarrow \langle t, e \rangle$
11:             **if** $n[e] \notin S$ **then**
12:                 Enqueue($S, n[e]$)
13:             **end if**
14:         **end if**
15:         $g_1(t, n[e]) \leftarrow g'$
16:     **end for**
17:     **if** $s \notin S'$ and $\{e | e \in E_1(s), i[e] \neq \varepsilon\} \neq \emptyset$ **then**
18:         Enqueue(S', s)
19:     **end if**
20: **end while**
21: **return** S'

---

weights up to the $(m - 1)$-th WFST, g is the current co-hypothesis list associated with the hypothesis that has reached state $n[e]$ at frame t in the $(m - 1)$-th WFST, and e is a composed transition up to the $(m - 1)$-th WFSTs. And the function returns the rescored weight $\hat{\lambda}$ and the co-hypothesis list $\hat{g}$ for the $(m - 1)$-th WFST.

On line 1, the function rescore() first checks whether the output label $o[e]$ is epsilon or not. If $o[e] \neq \varepsilon$, then it updates the co-hypothesis list and computes the rescored weight in lines 2–24. If $o[e] = \varepsilon$, it checks if the current co-hypothesis list g needs to be merged with another at the destination state $n[e]$ on line 26, i.e., it checks whether there is already another (co)hypothesis at state $n[e]$ and frame t and also checks whether its co-hypothesis list is identical to g. If it is necessary to merge the co-hypothesis lists, function merge() is called on line 27, or else nothing is done (α and g are simply copied to $\hat{\alpha}$ and $\hat{g}$ in lines 29 and 30 to return them unaltered).

The main rescoring steps are performed in lines 2–24. First we obtain the existing co-hypothesis list at the destination state $n[e]$ and frame t using a function update_cohyps() and assign the list to $\hat{g}$. The function returns $g_{m-1}(t, n[e])$ but if the co-hypotheses in $g_{m-1}(t, n[e])$ have not proceeded to frame t, their cumulative weights and back pointers need to be updated. We describe update_cohyps() in detail later.

Algorithm 29 transition_with_input($\mathbf{E}, S, A, X, t$)

1: **for** each $s \in S$ **do**
2:   **for** each $e \in E_1(s)$ such that $i[e] \neq \varepsilon$ **do**
3:     $\alpha(t-1, e, i_e) \leftarrow \alpha_1(t-1, s) \otimes w[e]$
4:     $b(t-1, e, i_e) \leftarrow t-1$
5:     $g(t-1, e, i_e) \leftarrow g_1(t-1, s)$
6:     **if** $\langle e, i_e \rangle \notin A$ **then**
7:       $A \leftarrow A \cup \{\langle e, i_e \rangle\}$
8:     **end if**
9:   **end for**
10: **end for**
11: $S' \leftarrow A' \leftarrow \emptyset$
12: **while** $A \neq \emptyset$ **do**
13:   $\langle e, j \rangle \leftarrow \text{Head}(A)$
14:   $\text{Dequeue}(A)$
15:   **for** each $k \in \text{Adj}(j)$ such that $k \neq f_e$ **do**
16:     $\alpha' \leftarrow \alpha(t-1, e, j) \otimes \omega(x[t], k|i[e], j)$
17:     **if** $\alpha(t, e, k) \otimes \alpha' = \alpha'$ **then**
18:       $\alpha_1(t, e, k) \leftarrow \alpha'$
19:       $b(t, e, k) \leftarrow b(t-1, e, j)$
20:       $g(t, e, k) \leftarrow g(t-1, e, j)$
21:       **if** $\langle e, k \rangle \notin A'$ **then**
22:         $\text{Enqueue}(A', \langle e, k \rangle)$
23:       **end if**
24:     **end if**
25:   **end for**
26: **end while**
27: **for** each $\langle e, k \rangle \in A'$ such that $f_e \in \text{Adj}(k)$ **do**
28:   $\alpha' \leftarrow \alpha(t, e, k) \otimes \omega(\varepsilon, f_e|i[e], k)$
29:   $\langle \alpha', g' \rangle \leftarrow \text{rescore}(\mathbf{E}, \alpha', g(t, e, k), e, t, 2)$
30:   **if** $\alpha_1(t, n[e]) \oplus \alpha' = \alpha'$ **then**
31:     $\alpha_1(t, n[e]) \leftarrow \alpha'$
32:     $B_1(t, n[e]) \leftarrow \langle b(t, e, k), e \rangle$
33:     **if** $n[e] \notin S'$ **then**
34:       $\text{Enqueue}(S', n[e])$
35:     **end if**
36:   **end if**
37:   $g_1(t, n[e]) \leftarrow g'$
38: **end for**
39: **return** $\langle A', S' \rangle$

Algorithm 30 rescore($\mathbf{E}, \alpha, g, e, t, m$)

1: **if** $o[e] \neq \varepsilon$ **then**
2:    $\hat{g} \leftarrow$ update_cohyps($n[e], t, m$)
3:    $\beta \leftarrow \left\{ \bigoplus_{\langle \tau, q \rangle \in g} \alpha_m(\tau, q) \right\}^{-1} \otimes \alpha$
4:    **for** each $\langle \tau, q \rangle \in g$ **do**
5:      $(s', s'') \leftarrow q$
6:      **for** each $r \in E_m(s,'' o[e])$ **do**
7:        $\alpha' \leftarrow \alpha_m(\tau, q) \otimes \beta \otimes w[r]$
8:        $r' \leftarrow \langle q, \varepsilon, o[r], w[r], (n[e], n[r]) \rangle$
9:        **if** $m < M$ **then**
10:          $\langle \alpha', g' \rangle \leftarrow$ rescore($\mathbf{E}, \alpha', g_m(\tau, q), r', t, m + 1$)
11:          $g_m(t, (n[e], n[r])) \leftarrow g'$
12:        **end if**
13:        **if** $\alpha_m(t, (n[e], n[r])) \oplus \alpha' = \alpha'$ **then**
14:          $\alpha_m(t, (n[e], n[r])) \leftarrow \alpha'$
15:          $B_m(t, ([n[e], n[r])) \leftarrow \langle \tau, r' \rangle$
16:        **end if**
17:        **if** $\hat{g} \notin \langle t, (n[e], n[r]) \rangle$ **then**
18:          Enqueue($\hat{g}, \langle t, (n[e], n[r]) \rangle$)
19:        **end if**
20:        **if** $\hat{\alpha} \oplus \alpha' = \alpha'$ **then**
21:          $\hat{\alpha} \leftarrow \alpha'$
22:        **end if**
23:      **end for**
24:    **end for**
25: **else**
26:    **if** $\alpha_{m-1}(t, n[e]) \neq \bar{0}$ and $g_{m-1}(t, n[e]) \neq g$ **then**
27:      $\hat{g} \leftarrow$ merge($\alpha, g, n[e], t, m$)
28:    **else**
29:      $\hat{g} \leftarrow g$
30:    **end if**
31: **end if**
32: **return** $\langle \hat{\alpha}, \hat{g} \rangle$

On line 3, β is obtained as a cumulative weight through transitions from the last time the co-hypotheses were extended. α should have been computed as the $\otimes$-product of the best co-hypothesis weight in g, i.e., $\bigoplus_{\langle \tau, q \rangle \in g} \alpha_m(\tau, q)$, and the cumulative weight β. Hence, β can be obtained as line 3. In lines 4–24, each co-hypothesis in g is extended with symbol $o[e]$. In line 5, since state q is a composed state, state s'' is extracted from q, which corresponds to the state the co-hypothesis has reached in the m-th WFST. In lines 6–23, the co-hypothesis is extended by adding each transition r that accepts $o[e]$ outgoing from state s''. The weight of the new co-hypothesis, α', is obtained on line 7 as $\alpha_m(\tau, q) \otimes \beta \otimes w[r]$. The composed transition r' is also obtained to keep track of the co-hypothesis at line 8. These are assigned to $\alpha_m(t, (n[e], n[r]))$ and $B_m(t, (n[e], n[r]))$ in lines 14 and 15, respectively, if α' is better than an already assigned weight to $\alpha_m(t, (n[e], n[r]))$.

The function rescore() is called recursively if $m < M$. On line 10, α' is replaced with the rescored weight using the $(m + 1)$-th and later WFSTs. The co-hypothesis list $g_m(t, (n[e], n[r]))$ is also replaced with the extended co-hypothesis list g' on line 11.

In lines 17–19, new co-hypothesis list $\hat{g}$ is constructed by inserting each co-hypothesis represented as a pair $\langle t, (n[e], n[r]) \rangle$. The pair means that the co-hypothesis has reached a composed state $(n[e], n[r])$ at frame t. In lines 20–22, the rescored weight $\hat{\alpha}$ is computed as the best co-hypothesis weight in $\hat{g}$. $\hat{\alpha}$ and $\hat{g}$ are finally returned to the invoker on line 32.

Note that we assume here for simplicity that the 2nd to m-th WFSTs are input-epsilon-free transducers, i.e., they have no epsilon input label in each transition. To use WFSTs with epsilon input labels, we can extend the function so that, for example, all possible epsilon transitions are made using a local queue of states for each m-th WFST.

Functions update_cohyps() and merge() used in rescore() are given in Algorithms 31 and 32. Function update_cohyps() first checks if a co-hypothesis in list $g_{m-1}(t, s)$ has already reached

---

Algorithm 31 update_cohyps(s, t, m)

1: **if** $\exists \langle \tau, (s', s'') \rangle \in g_{m-1}(t, s), \tau = t$ and $s' = s$ **then**
2:    **return** $g_{m-1}(t, s)$
3: **else**
4:    $\hat{g} \leftarrow \emptyset$
5:    $\beta \leftarrow \left\{ \bigoplus_{\langle \tau, q \rangle \in g_{m-1}(t, s)} \alpha_m(\tau, q) \right\}^{-1} \otimes \alpha_{m-1}(t, s)$
6:    **for** each $\langle \tau, q \rangle \in g_{m-1}(t, s)$ **do**
7:       $\langle s', s'' \rangle \leftarrow q$
8:       $\alpha_m(t, (s, s'')) \leftarrow \alpha_m(\tau, q) \otimes \beta$
9:       $B_m(t, (s, s'')) \leftarrow B_m(\tau, q)$
10:      Enqueue($\hat{g}, \langle t, (s, s'') \rangle$)
11:   **end for**
12:   **return** $\hat{g}$
13: **end if**

state s at frame t on line 1. If this condition is satisfied, it simply returns $g_{m-1}(t, s)$ unaltered. If the condition is not satisfied, each co-hypothesis weight is updated with β obtained on line 5, and assigned to $\alpha_m(t, (s, s''))$ on line 8. The back pointer is also updated. On line 10, the updated co-hypothesis is stored in list $\hat{g}$.

Algorithm 32 merge(α, g, s, t, m)

1: $\hat{g} \leftarrow$ update_cohyps(s, t, m)

2: $\beta \leftarrow \left\{ \bigoplus_{\langle \tau, q \rangle \in g} \alpha_m(\tau, q) \right\}^{-1} \otimes \alpha$

3: **for** each $\langle \tau, q \rangle \in g$ **do**

4:      $(s', s'') \leftarrow q$

5:      $\alpha' \leftarrow \alpha_m(\tau, q) \otimes \beta$

6:      **if** $m < M$ **then**

7:          $g' \leftarrow g_m(\tau, q)$

8:          **if** $\alpha_m(t, (s, s'')) \neq \bar{0}$ and $g_m(t, (s, s'')) \neq g'$ **then**

9:              $g' \leftarrow$ merge($\alpha', g', (s, s''), t, m + 1$)

10:         **end if**

11:         $g_m(t, (s, s'')) \leftarrow g'$

12:     **end if**

13:     **if** $\alpha_m(t, (s, s'')) \oplus \alpha' = \alpha'$ **then**

14:         $\alpha_m(t, (s, s'')) \leftarrow \alpha'$

15:         $B_m(t, (s, s'')) \leftarrow B_m(\tau, q)$

16:     **end if**

17:     **if** $\hat{g} \in \langle t, (s, s'') \rangle$ **then**

18:         Enqueue($\hat{g}, \langle t, (s, s'') \rangle$)

19:     **end if**

20: **end for**

21: **return** $\hat{g}$

Function merge() merges two co-hypothesis lists. One is $\hat{g}$, which equals $g_{m-1}(t, s)$, and the other is g given by the invoker. On line 1, $\hat{g}$ is updated as necessary. On line 2, β is obtained in a similar way to rescore(). In lines 3–20, each co-hypothesis in g is inserted in $\hat{g}$. The co-hypothesis weight is calculated with $\alpha_m(\tau, q)$ and β on line 5. This function is also called recursively if $m < M$ in lines 6–12 to merge all the associated co-hypothesis lists. The cumulative weight and the back pointer are assigned in lines 13–16, and the merged co-hypothesis list is constructed in lines 17–19. The merged list is finally returned on line 21.

Algorithm 33 shows the pseudo-code of final_transition() for the rescoring approach. This is also similar to that for decoding with a single WFST in Algorithm 22. Function rescore() is called in line 7 to rescore the hypotheses in the final epsilon transitions. On line 19, final_rescore() is included, which is used to find the best co-hypothesis that has reached a composed state of final states in the M WFSTs.

Algorithm 33 final_transition($\mathbf{E}, \mathbf{F}, \mathbf{R}, S, T$)

1: $\hat{\alpha} \leftarrow \bar{0}$
2: **while** $S \neq \emptyset$ **do**
3:    $s \leftarrow \text{Head}(S)$
4:    $\text{Dequeue}(S)$
5:    **for** each $e \in E_1(s, \varepsilon)$ **do**
6:       $\alpha' \leftarrow \alpha_1(T, s) \otimes w[e]$
7:       $\langle \alpha', g' \rangle \leftarrow \text{rescore}(\mathbf{E}, \alpha', g_1(T, s), e, T, 2)$
8:       **if** $\alpha_1(T, n[e]) \oplus \alpha' = \alpha'$ **then**
9:          $\alpha_1(T, n[e]) \leftarrow \alpha'$
10:         $B_1(T, n[e]) \leftarrow \langle T, e \rangle$
11:         **if** $n[e] \notin S$ **then**
12:            $\text{Enqueue}(S, n[e])$
13:         **end if**
14:       **end if**
15:       $g_1(T, n[e]) \leftarrow g'$
16:    **end for**
17:    **if** $s \in F_1$ **then**
18:       $\alpha' \leftarrow \alpha_1(T, s) \otimes \rho_1(s)$
19:       $\langle \alpha', q' \rangle \leftarrow \text{final\_rescore}(\mathbf{F}, \mathbf{R}, \alpha', s, T, 2)$
20:       **if** $\hat{\alpha} \oplus \alpha' = \alpha'$ **then**
21:          $\hat{\alpha} \leftarrow \alpha'$
22:          $\hat{B} \leftarrow B_M(T, q')$
23:       **end if**
24:    **end if**
25: **end while**
26: **return** $\hat{B}$

Function final_rescore() is shown in Algorithm 34. The best co-hypothesis and its final state in co-hypothesis list $\hat{g}$ ($= g_m(t, s)$) is obtained in lines 3–15. If state s'', which the co-hypothesis has reached in the m-th WFST, is a final state in F_m, the co-hypothesis weight is multiplied by the final weight $\rho_m(s'')$ on line 6. The final weight is also rescored recursively up to the M-th WFST in lines 7–9. The best co-hypothesis weight and its final state are memorized in $\hat{\alpha}$ and $\hat{q}$, which are finally returned to the invoker on line 16.

Although sub-codes prune() and backtrack() used in the rescoring approach are not presented in this section, we can use those for a single WFST as they are. In addition, we may employ additional

---

Algorithm 34 final_rescore($\mathbf{F}, \mathbf{R}, \alpha, s, t, m$)

1: $\hat{g} \leftarrow$ update_cohyps(s, t, m)
2: $\hat{\alpha} \leftarrow \bar{0}$
3: **for** each $\langle \tau, q \rangle \in \hat{g}$ **do**
4:    $(s', s'') \leftarrow q$
5:    **if** $s'' \in F_m$ **then**
6:       $\alpha' \leftarrow \alpha_m(\tau, q) \otimes \rho_m(s'')$
7:       **if** $m < M$ **then**
8:          $\langle \alpha', q' \rangle \leftarrow$ final_rescore($\mathbf{F}, \mathbf{R}, \alpha', (s, s''), t, m + 1$)
9:       **end if**
10:      **if** $\hat{\alpha} \oplus \alpha' = \alpha'$ **then**
11:         $\hat{\alpha} \leftarrow \alpha'$
12:         $\hat{q} \leftarrow q'$
13:      **end if**
14:    **end if**
15: **end for**
16: **return** $\langle \hat{\alpha}, \hat{q} \rangle$

---

pruning for co-hypotheses to reduce the computation needed for rescoring. For example, we may limit the size of the list to a predefined number.

5.5.4 APPROXIMATION IN DECODING

The on-the-fly rescoring approach shown in Algorithms 25–34 does not necessarily ensure that the best hypothesis is found. This is usually not a big problem but this property is important. In speech recognition, the first WFST may have an HMM state sequence on each transition as its input label if the HMM WFST H is represented as in Fig. 4.4. Nevertheless, we can use a factorization operation to replace chained transitions with one transition. The input label may result in a longer HMM state sequence.

HMM-level state transitions in one transition are handled in Algorithm 29. In the Viterbi computation for the HMM-level state transitions, we avoid merging the co-hypothesis lists when one hypothesis encounters another in an HMM state because frequent list merging may impose a certain overhead although each merging process involves little computation. However, this simplification requires an assumption to ensure that the best (co-)hypothesis is found.

Figure 5.13 shows an example of an error that occurs when the assumption is not satisfied. A trellis space formed by part of a WFST and a time axis can be seen in the figure. The WFST-like graph shown on the Y-axis is a network that was actually searched by the decoder. The network comprises four WFST states (the numbered large nodes) and 11 HMM states (the small nodes labeled by HMM-state Ids as "Sk" where k is an integer).

There are three Viterbi paths representing hypotheses in the trellis space. At time t_1, one hypothesis encounters another, and only the better hypothesis survives and proceeds following t_1. Then the two co-hypothesis lists attached to the hypotheses are merged into one list that is delivered to the surviving hypothesis. However, at time t_2, the co-hypothesis list attached to the worse hypothesis represented with a dashed line is lost since we avoid merging the co-hypothesis lists when a hypothesis encounters another in an HMM state. Accordingly, search errors will occur as a result of the loss of the co-hypothesis lists in an HMM state if the co-hypothesis that will become the best complete co-hypothesis is included in the lost list. The loss of such co-hypotheses also affects the weights of the succeeding hypotheses. To ensure both the best hypothesis and co-hypothesis are retained, the boundary time between S4 and S9 (i.e., WFST state 3) must be equal to that between S8 and S9 for all the hypotheses.

However, as mentioned in [SYM+96] triphones yield a good assumption. If cross-word triphones are used, all transitions to a WFST state come from HMM states associated with a unique preceding phone. In that case, when hypotheses ending with the same phone meet in a WFST state during decoding, the boundary time between a phone and a succeeding phone tends to be equal on the Viterbi paths. This kind of assumption is also used to generate word lattices in the WFST framework [LPR99].

In addition, even if the boundary times are not the same, critical errors rarely occur because the same or similar co-hypotheses with the same output symbol sequence are usually included in both surviving and lost lists. But the cumulative weight may change slightly due to a declination of the time alignment. Since the time alignment is decided based on the first WFST as well as the two pass search [LPR99], it can be different from that derived by the composed WFST. In Fig. 5.12, for example, the boundary times between S4 and S6 and between S5 and S6 necessarily share a common time decided by the first WFST, while they can be different from each other in Fig. 5.11. Hence, when using the proposed method, a declination of the time alignment can occur even if the correct output sequence is derived.

Although the preceding phone is not necessarily unique since the WFST is actually optimized up to the shared HMM states, we can say that *phone-pair approximation* can be roughly assumed. The phone-pair approximation assumes that the best starting time for a phone depends solely on the preceding phone rather than on the entire preceding phone sequence. In the example in Fig. 5.13, since HMM states S4 and S8 belong to the same center phone or similar phones, the boundary times for S9 will be the same or similar in terms of the time alignments of the hypotheses. If this assumption is satisfied, the rescoring approach ensures that the best (co-)hypothesis is found. Even if the assumption is not satisfied, the error rarely militates against the recognition result since the cumulative weight does not change significantly with small differences in time alignment.

5.5.5   COMPARISON WITH LOOK-AHEAD COMPOSITION

Dixon et al. experimentally compared the two dynamic decoding approaches, i.e., look-ahead composition and on-the-fly rescoring [DHK12]. They reported that a slight degradation of word accuracy

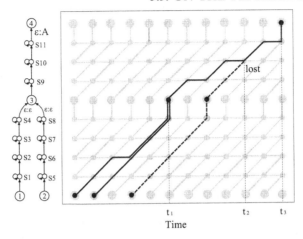

Figure 5.13: Example of approximation error: There are three hypotheses in a trellis space organized with time on the X-axis and a search network on the Y-axis. This example shows that one of the hypotheses can be lost with its co-hypothesis list due to the mismatch in the time alignment from the other hypotheses.

was observed when applying the on-the-fly rescoring approach with a combination of a unigram-based WFST $HCLG_1$ and the corresponding trigram WFST $G_{3/1}$. However, the degradation disappeared when they used a bigram-based WFST $HCLG_2$ and the trigram WFST $G_{3/2}$. This is because the approximation was alleviated by using a bigram language model for the first WFST.

Finally, they concluded that these two approaches had the same performance in terms of speed and accuracy. In addition, they reported that the rescoring approach worked with less memory consumption than that of the look-ahead composition approach. This seems to be an advantage of the rescoring approach, but we should note that the look-ahead composition has another advantage that no approximation is used in the decoding process and no incremental language models need to be handled in the construction process of WFSTs.

CHAPTER 6

Summary and Perspective

In this book, we introduced speech recognition algorithms based on Weighted Finite-State Transducers (WFSTs). First, we briefly described the fundamentals of large-vocabulary continuous-speech recognition (LVCSR). Specifically, we focused on the decoding problem in LVCSR and discussed why the WFST-based approach improved search efficiency in the decoding process. We also presented concrete algorithms for constructing a fully composed WFST and decoding with the WFST. Finally, we described two dynamic decoders with on-the-fly composition of multiple WFSTs, which are more tractable for practical speech applications than the basic approach with a fully composed WFST.

As we mentioned in the Introduction, the success of the WFST approach in speech recognition could result both from its enhanced performance in terms of speed and its elegance, which opened new vistas in the field. The use of this general approach is not limited to decoding in speech recognition. It covers a wide area from classical text processing to integrated applications that handle various sequential inputs. Besides decoding, it has recently been used for training acoustic, pronunciation, and language models for speech recognition. Other applications such as spoken language understanding (SLU), dialog management (DM), speech synthesis, spoken term detection (STD), and machine translation have also been achieved using WFSTs. Moreover, different transduction processes can be combined into one process such as speech summarization and speech translation, where WFSTs for speech recognition and the following language processing are combined with the composition operation and the composed WFST is used to obtain the summarization or translation result directly from speech. The approach is also extended to accept a multi-modal input based on multi-tape WFSTs. Here we describe a brief overview of recent advances in WFST-based speech applications, and conclude this book.

6.1 REALIZATION OF ADVANCED SPEECH RECOGNITION TECHNIQUES USING WFSTS

Chapter 4 described the basic WFST-based LVCSR procedure, which utilizes triphone HMMs and a word n-gram language model. However, there are other speech recognition techniques that further improve the recognition accuracy. Although most techniques can be used in the WFST framework, some need additional steps if they are to be used effectively. This section introduces several examples of advanced speech recognition techniques using WFSTs.

6.1.1   WFSTS FOR EXTENDED LANGUAGE MODELS

A class-based n-gram model [BdM$^+$92] is sometimes chosen according to the application domain and the available training data. With the class model, the occurrence probability of w_n given a history $w_1, \ldots, w_{n-1}$ is computed as

$$P(w_n|C(w_n))P(C(w_n)|C(w_1), \ldots, C(w_{n-1})), \tag{6.1}$$

where $C(w)$ denotes the class the word w belongs to. For good language modeling, the classes need to be designed to contain similar words that we hope occur in the same context.

The effect of the class models is to interpolate the probabilities of unobserved word n-grams in training data. This means that, for an unobserved n-gram, $u_1, \ldots, u_n$, we can obtain a good prediction of the probability, i.e., it is interpolated with the probability of the n-class sequence, $P(C(u_n)|C(u_1), \ldots, C(u_{n-1}))$. This model is also convenient when we add a new word v to the speech recognizer, because we only need to assign one of the classes to the word and set an arbitrary probability for $P(v|C(v))$.

Allauzen et al. described how to construct a WFST for a class n-gram model [AMR03], which can be composed as

$$G = \pi_1(M \circ G_C), \tag{6.2}$$

where M denotes a word-to-class mapping WFST with a weight of $P(w|C(w))$ on each transition, and G_C denotes a WFST of an n-gram model of class labels, which gives a weight calculated based on probability $P(C(w_n)|C(w_1) \ldots C(w_{n-1}))$. $\pi_1()$ is a projection operator that overwrites the output label (i.e., class) of each transition with its input label (i.e., word). However, G can be very large because each transition labeled with a class in G_C is multiplied by transitions of words belonging to the class in M. Hence, on-the-fly composition and projection may be employed as in [HN05] to reduce the memory consumption.

Recently, discriminative language models (DLMs) have been used to improve speech recognition accuracy [RSCJ04]. With a DLM, the speech recognition in Eq. (2.3) is extended to

$$\hat{W} = \underset{W \in \mathcal{W}}{\text{argmax}} \left\{ \log p(O|W)P(W) + \mathbf{a} \cdot \mathbf{f}(W) \right\}, \tag{6.3}$$

where the first term corresponds to a traditional speech recognition score calculated as the acoustic likelihood $p(O|W)$ times the language probability $P(W)$ for a hypothesized word sequence W given input speech O. The additional term $\mathbf{a} \cdot \mathbf{f}(W)$ represents the DLM score, which is calculated as the inner product of the weight vector $\mathbf{a}$ and the feature vector $\mathbf{f}(W)$ extracted from W. $\mathbf{a}$ is trained to reduce the recognition errors in advance.

In this approach, we can utilize different linguistic information in W such as word n-grams, part-of-speech (POS) n-grams, and the pronunciation of each word. With such features and a discriminatively trained weight vector, the speech recognition score is modified, and a recognition result with fewer errors can be selected.

A DLM can also be represented as a set of WFSTs. For example, a WFST representing a DLM with only word n-gram features has a similar structure to that of a word n-gram language model, where each transition corresponds to an arbitrary n-gram feature and has the weight for the n-gram feature in **a**. When we use POS n-gram features in addition to word n-gram features, WFST D of a DLM can be composed as

$$D = \pi_1(D_W \circ M_{WP} \circ D_P), \tag{6.4}$$

where D_W, M_{WP}, and D_P denote WFSTs for word n-gram features, the mapping of each word to its POS label, and POS n-gram features, respectively. These WFSTs are available in addition to the standard language model by using on-the-fly composition [OHIN11].

It is well known that a combination of language models in different domains or tasks is an effective way to improve the recognition accuracy when we cannot obtain enough training data for the target task [JMSS91, IOR94]. In the WFST approach, the combination of multiple language models is performed by a union or intersection operation of the language model WFSTs. The union results in a linear combination of probabilities while the intersection results in a log-linear combination [LGHW10]. Furthermore, language models over different speaking styles or even different languages can be combined using the WFST framework. In [HWM03a], written-style and spoken-style models are combined by introducing a style conversion WFST. In [JOIF09], an Icelandic model is combined with an English model by introducing a translation WFST to compensate for the resource deficiency when training the Icelandic model.

6.1.2   DYNAMIC GRAMMARS BASED ON WFSTS

As an extension of the WFST approach, a dynamic grammar-based speech recognition technique using WFSTs was proposed [JSHS03], in which the dynamic expansion of states and transitions was performed when the path hypothesis entered a transition with a non-terminal symbol during decoding. This is no longer a finite-state automaton, but it has become a push-down automaton. A technique called *splicing* is used to dynamically insert another WFST at the transition with the non-terminal label to which the inserted WFST is assigned. This approach is useful for changing the vocabulary (e.g., adding or deleting a word entry for a semantic class) [Het05], while the original WFST approach needs to reconstruct the fully composed WFST. In this case, we prepare a WFST for each word class such as *PERSON_NAME*, *TRAIN_STATION*, and *ORGANIZATION_NAME*. Since these WFSTs are expanded dynamically during decoding, reconstruction of the whole WFST is not required even if we modify such component WFSTs. The push-down automaton can also be used to represent a context free grammar (CFG) that can be used as a language model for speech recognition.

6.1.3   WIDE-CONTEXT-DEPENDENT HMMS

We can utilize wide-context-dependent acoustic units beyond triphones, called *polyphones*, to improve recognition accuracy [BdSG91, KLNB95]. For example, quinphones consider the context dependency of the two preceding and two succeeding phones. The HMM parameters can be estimated based on such a wide context dependency. But with the WFST approach, it is very expensive to construct a context-dependency WFST C for polyphones because the WFST needs $|P|^{K-1}$ states and $|P|^K$ transitions, where K denotes the order of polyphones and $|P|$ is the number of basic phone units, which is usually over 40, e.g., a quinphone WFST needs approximately $40^4(= 2.56 \times 10^6)$ states and $40^5(= 10^8)$ transitions. In reality, most transitions of such a polyphone WFST accept the same HMM state sequence because the number of shared states is empirically much smaller than the number of unique polyphones. Hence, given an HMM WFST H, the size of $HC = det(H \circ C)$ is actually not very large. Some techniques for directly constructing HC have been proposed [SH05, SM06] to avoid constructing C for the polyphones. The HC can be composed with L and G according to the standard process of WFST construction for speech recognition.

6.1.4   EXTENSION OF WFSTS FOR MULTI-MODAL INPUTS

As an extension of the WFST approach, multi-tape WFSTs are used to solve the problem of multi-modal speech recognition. A multi-tape WFST can accept multi-stream inputs by extending the definition of the WFST so that it has multiple input labels on each transition. In [HSG06], a multi-tape WFST is applied to audio-visual speech recognition using both speech and visual (lip movement) input. This work has proposed an efficient decoding algorithm with control of the degree of asynchrony between input streams. The technique was also used to integrate frame-based and segment-based acoustic models.

6.1.5   USE OF WFSTS FOR LEARNING

In general, WFST-based speech recognition is known as an efficient decoding approach. However, the WFST framework has also been used to learn underlying models or the WFST itself because of its utility.

In the discriminative training of acoustic models, lattices are used to represent a set of competing hypotheses generated with the current acoustic model. The expected loss such as the expected number of phone errors [PW02] is estimated using the lattice and the reference, and the parameters of the models are updated to reduce the expected loss. By representing the lattices and references with WFSTs, we can easily manipulate the importance of each hypothesis and can also exclude some symbol matches from the loss function calculation [MHLR+07]. Furthermore, some algorithms have been proposed for obtaining the expected loss using WFST operations [HSN09, HHR+12, ZLCJ12].

On the other hand, an EM algorithm can be employed to estimate the transition weights of WFSTs [SH02]. The maximum likelihood estimates of the transition weights are obtained using a set of input/output symbol sequence pairs in a training corpus. In E-step, we can obtain the expected count for each transition by accumulating posterior probabilities over the paths that accept the sequence pairs. In M-step, we can obtain the maximum likelihood estimates of transition weights by normalizing the expected counts so that the total weights of all transitions leaving each state is 1. The resulting WFST is a joint probability model of the input and output sequences. The WFST can be converted into a conditional probability model using WFST operations. In [Eis02], a method is presented for training WFSTs using an expectation semiring.

Furthermore, a WFST can be trained discriminatively since it can be considered a log-linear model such as a conditional random field (CRF) that classifies sequential patterns [WHMN10, KWHN12]. Thus, several training techniques are available, where we assume that each transition of the WFST corresponds to a feature that matches an input symbol and is weighted by the transition weight in the log domain.

The CRF-based approach has already been incorporated in speech recognition as Hidden CRF and Segmental CRF [ZN09] independently of WFSTs. The advantage of this approach is that it realizes the optimal integration of basic features such as acoustic and language model scores and extended features such as phone duration [LS10] and acoustic distance by template matching [HZN09] in a discriminative criterion.

Accordingly, we can extend the WFST to improve the speech recognition accuracy by incorporating additional features. For example, in [WHMN10], the authors introduced acoustic observations at each transition and the occurrence count of the transition as such features. In this case, a weight vector is associated with each transition from which the feature vector is extracted. The inner product of the weight and feature vectors is used as the transition weight during decoding. Hence, we do not have to change the algorithm of the decoder.

6.2   INTEGRATION OF SPEECH AND LANGUAGE PROCESSING

One versatile feature of the WFST framework is that different transduction processes in a cascade can be integrated into one transduction process. As mentioned in Section 4, transductions in speech recognition can be combined into a single WFST. From this point of view, the speech recognition process can be composed further with another process that handles speech recognition results. For example, the speech-input machine translation approach [Cas01] integrates speech recognition and machine translation into one transduction process.

The translation of a source language W into a target language can be formulated as the search for a word sequence $\hat{T}$ from a target language, such that

$$\hat{T} = \underset{T}{\operatorname{argmax}} P(T|W) \tag{6.5}$$

$$= \underset{T}{\operatorname{argmax}} P(W|T)P(T), \tag{6.6}$$

where $P(W|T)$ is a translation model between W and T and $P(T)$ is a language model of the target language. If the source language is speech O, i.e., a speech input case, the translation can be formulated as the search for $\hat{T}$ such that

$$\hat{T} = \underset{T}{\operatorname{argmax}} P(T|O) \tag{6.7}$$

$$= \underset{T}{\operatorname{argmax}} \sum_{W} P(O|W,T)P(W|T)P(T) \tag{6.8}$$

$$\simeq \underset{T}{\operatorname{argmax}} \max_{W} P(O|W)P(W|T)P(T). \tag{6.9}$$

If we can construct a WFST for each probabilistic model, the decoding process can be achieved by the WFST approach.

For example, we assume the translation probability $P(W|T)$ as

$$P(W|T) \approx P_G(W)\delta_S(W,T), \tag{6.10}$$

where $P_G(W)$ is a prior probability of W given by a language model for speech recognition, and $\delta_S(W,T)$ takes a binary 0 or 1 value depending on whether or not it is possible to substitute W with T, which is given by a set of substitution rules of word sequences. The substitution function $\delta_S(W,T)$ can be expressed as a WFST [HWM03b].

Let S be a WFST of $\delta_S(W,T)$, and D be a WFST of a language model of the target language. The integrated WFST for speech translation can be composed as

$$Z = H \circ C \circ L \circ G \circ S \circ D. \tag{6.11}$$

The speech translation process is illustrated by the cascade in Fig. 6.1. The advantage of this in-

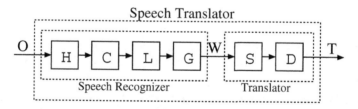

Figure 6.1: Cascade of speech-input translation

tegration is that higher level knowledge sources are available from the beginning of the decoding, and finally the best hypothesis can be found by considering all the knowledge sources. This general framework is also applicable to other types of spoken language processing, such as speech paraphrasing [HWM03b], speech summarization [HHM03], and spoken language understanding, which generates a sequence of concept or semantic tags representing the spoken content.

On the other hand, the lattice-based approach is also used for integrating speech recognition and language processing without composing all the WFSTs. General speech recognizers can output both a single word sequence and a word lattice representing a set of the most likely hypotheses for the input utterance. The language processor can work better using the lattice than using a single hypothesis because it may be unable to handle the single hypothesis that includes recognition errors. If the lattice is available, the processor can choose a better hypothesis from the lattice, which potentially includes fewer recognition errors.

The WFST approach is effective if the language processing is also designed as a WFST. Let the WFST for language processing be F and the word lattice be L, which is also represented as a WFST. The final result of the language processing is obtained as

$$R = \text{bestpath}(L \circ F), \qquad (6.12)$$

where $\text{bestpath}(\cdot)$ denotes an operation that returns a minimally weighted path in the WFST.

The word lattices are also used for spoken utterance retrieval and classification problems. In [AM04], a general indexation method is presented in which a set of word lattices over all the spoken documents are converted into an index for spoken utterance retrieval. The index is constructed as a large WFST that can be seen as a suffix automaton including all the lattices. The index construction and the search are achieved by using WFST operations efficiently. In the search step, the composition operation for the index and a query, which is also represented as a WFST, results in a WFST representing a list of the matched utterance Ids.

In [CHM04], word lattices are employed in spoken utterance classification problems, where each lattice is used as an input for a classifier such as a support vector machine (SVM). Rational kernels defined over WFSTs are used to measure similarities between the lattices. The SVM was trained using a set of lattices labeled with the corresponding category. It has been shown that the use of lattices outperforms the accuracy achieved when only using a single hypothesis in call routing and spoken dialog tasks. The paper has shown that the WFSTs can be applied effectively to the classification problem by using a rational kernel.

In [BJ09], a multi-tape WFST is used as a multi-modal grammar that yields an effective mechanism for quickly creating integration and understanding capabilities for interactive systems supporting the simultaneous use of multiple input modalities. To understand a multi-modal input as a single concept, composition of a multi-modal grammar and recognition lattices of different modalities is efficiently performed, where each lattice is obtained by an independent recognizer for each modality such as speech input and freehand pen input on the map. With this approach, the most consistent understanding can be derived by considering both inputs.

6.3   OTHER SPEECH APPLICATIONS USING WFSTS

There are many WFST-based applications besides speech recognition. WFSTs are used in speech synthesis to perform several transductions [Spr96] such as grapheme-to-phoneme conversion [CTOV02], unit selection and prosody prediction so that natural speech is synthesized [BO01].

In [BO02], language generation and speech synthesis are efficiently integrated by using WFSTs for spoken dialog systems. Specifically, multiple wordings of the response and multiple prosodic realizations of the different wordings are allowed. The choice of wording and prosodic structure are then jointly optimized with unit selection for waveform generation in speech synthesis. This is more effective for generating natural synthetic speech than cascaded synthesis.

In [HOM⁺08], a WFST representing possible dialog scenarios is used for dialog management (DM), where the WFST accepts a sequence of user's concepts and outputs a sequence of system actions. The WFST is then composed with the WFST for spoken language understanding (SLU) that transduces a speech recognition result into a concept sequence. The composed WFST can handle SLU and DM simultaneously, i.e., it converts a speech recognition result to possible concepts, and simultaneously changes the dialog state according to the scenario. This method achieves an appropriate understanding of a user's utterance depending on the dialog state, and can lead the dialog in the correct direction. Moreover, it can also be applied to statistical DM using the same platform [HOM⁺09].

6.4 CONCLUSION

In this book, we described speech recognition algorithms using WFSTs. We tried to explain this approach from its fundamentals to the concrete algorithms, and to include important techniques required for WFST-based speech recognition. We hope this book will prove useful for anyone who wants to understand, develop, and study the WFST approach in detail.

As we mentioned in the previous sections, currently WFSTs are widely used for various applications beyond speech recognition decoding. This noteworthy progress originates in their efficiency and generality. Moreover, we must not forget that the progress has been strongly supported by publicly available software for handling WFSTs. This software is well designed for performing various WFST operations efficiently. Table 6.1 summarizes popular software packages for WFST operations. Open source speech recognition decoders are also found in [MDMD⁺06, DDCTOSF07, PGB⁺11, RHL⁺11]. Currently, such software packages are being used by many researchers and developers that follow the WFST approach. Thus, the WFST framework is becoming a general-purpose tool for recognition, translation, and classification of sequential patterns. In the future, the WFST approach will continue to evolve, and the application area will expand further.

Table 6.1: Library of WFST operations

Name	Copyright	License	Feature	Reference
FSM Library	AT&T Corp.	Non-commercial use	Binary only	[MPR00] [FSM]
OpenFst Library	Google, Inc.	Apatch v2.0	C++ templates, Generalized composition	[ARS⁺07] [OPE]
MIT FST Toolkit	M.I.T.	BSD	Splicing, EM training	[Het04] [MIT]
RWTH FSA Toolkit	RWTH Aachen	GPL		[KN04] [RWT]

Bibliography

[ABCF95] G. Antoniol, F. Brugnara, M. Cettolo, and M. Frederico, "Language model representations for beam-search decoding," in *Proc. ICASSP*, 1995, pp. 588–591. DOI: 10.1109/ICASSP.1995.479666 33

[AHU74] A. V. Aho, J. E. Hopcroft, and J. D. Ullman, *The Design and Analysis of Computer Algorithms*. Addison-Wesley Publishing Company, 1974. 4

[AM04] C. Allauzen and M. Mohri, "General indexation of weighted automata—application to spoken utterance retrieval," in *Proc. HLT-NAACL*, 2004. 135

[AMR03] C. Allauzen, M. Mohri, and B. Roark, "Generalized algorithms for constructing statistical language models," in *Proc. ACL*, 2003, pp. 40–47. DOI: 10.3115/1075096.1075102 77, 130

[AMRR04] C. Allauzen, M. Mohri, M. Riley, and B. Roark, "A generalized construction of integrated speech recognition transducers," in *Proc. ICASSP*, vol. I, 2004, pp. 761–764. DOI: 10.1109/CASSP.2004.1326097 91

[ARS+07] C. Allauzen, M. Riley, J. Schalkwyk, W. Skut, and M. Mohri, "OpenFst: A general and efficient weighted finite-state transducer library," in *Proc. of CIAA*, 2007, pp. 11–23. DOI: 10.1007/978-3-540-76336-9_3 136

[ARS09] C. Allauzen, M. Riley, and J. Schalkwyk, "A generalized composition algorithm for weighted finite-state transducers," in *Proc. Interspeech*, 2009, pp. 1203–1206. 71, 91, 95, 99, 104, 105, 107

[ARS11] C. Allauzen, M. Riley, and J. Schalkwyk, "A filter-based algorithm for efficient composition of finite-state transducers," *International Journal of Foundations of Computer Science*, 2011. DOI: 10.1142/S0129054111009033 105, 106

[ASP91] S. Austin, R. Schwartz, and P. Placeway, "The forward-backward search algorithm," in *Proc. ICASSP*, vol. 1, 1991, pp. 697–700. DOI: 10.1109/ICASSP.1991.150435 4

[Aub02] X. L. Aubert, "An overview of decoding techniques for large vocabulary continuous speech recognition," *Computer Speech and Language*, vol. 16, pp. 89–114, 2002. DOI: 10.1006/csla.2001.0185 4

[BBC82] J. S. Bridle, M. D. Brown, and R. M. Chamberlain, "An algorithm for connected word recognition," in *Proc. ICASSP*, 1982, pp. 899–902. 3

[BBCF10] J. Berstel, L. Boasson, O. Carton, and I. Fagnot, "Minimization of automata," *CoRR*, vol. abs/1010.5318, 2010. 61

[BC95] F. Brugnara and M. Cettolo, "Improvements in tree-based language model representation," in *Proc. EUROSPEECH*, 1995, pp. 1797–1800. 33

[BD62] R. Bellman and S. Dreyfus, *Applied Dynamic Programming*. New Jersey: Princeton Univ. Press, 1962. 2, 16

[BdM+92] P. F. Brown, P. V. deSouza, R. L. Mercer, V. J. Della Pietra, and J. C. Lai, "Class-based n-gram models of natural language," *Computational Linguistics*, vol. 18(4), pp. 467–479, 1992. 130

[BdSG91] L. R. Bahl, P. V. de Souza, and P. S. Gopalakrishman, "Decision trees for phonological rules in continuous speech," in *Proc. ICASSP*, 1991, pp. 185–188. DOI: 10.1109/ICASSP.1991.150308 18, 131

[BJ09] S. Bangalore and M. Johnston, "Robust understanding in multimodal interfaces," *Computer Linguistics*, vol. 35, no. 3, pp. 345–397, 2009. DOI: 10.1162/coli.08-022-R2-06-26 135

[BJM83] L. R. Bahl, F. Jelinek, and R. L. Mercer, "Maximum likelihood approach to continuous speech recognition," *IEEE Transactions on Patten Analysis and Machine Intelligence*, vol. PAMI-5, no. 2, pp. 179–190, Mar. 1983. DOI: 10.1109/TPAMI.1983.4767370 9

[BL76] K. Booth and G. Lueker, "Testing for the consecutive ones property, interval graphs, and graph planarity using pq-tree algorithms," *Journal of Computer and System Sciences*, vol. 13, pp. 335–379, 1976. DOI: 10.1016/S0022-0000(76)80045-1 109

[BO01] I. Bulyko and M. Ostendorf, "Joint prosody prediction and unit selection for concatenative speech synthesis," in *Proc. ICASSP*, vol. 2, 2001, pp. 781–784. DOI: 10.1109/ICASSP.2001.941031 135

[BO02] I. Bulyko and M. Ostendorf, "Efficient integrated response generation from multiple targets using weighted finite state transducers," *Computer Speech and Language*, vol. 16(3-4), pp. 533–550, 2002. DOI: 10.1016/S0885-2308(02)00023-2 135

[Cas01] F. Casacuberta, "Finite-state transducers for speech-input translation," in *Proc. ASRU*, 2001, pp. 375–380. DOI: 10.1109/ASRU.2001.1034664 133

[CDD07] O. Cheng, J. Dines, and M. M. Doss, "A generalized dynamic composition algorithm of weighted finite state transducers for large vocabulary speech recognition," in *Proc. ICASSP*, 2007, pp. 348–351. DOI: 10.1109/ICASSP.2007.366920 95, 99

[CHM04] C. Cortes, P. Haffner, and M. Mohri, "Rational kernels: Theory and algorithms," *The Journal of Machine Learning Research*, vol. 5, pp. 1035–1062, 2004. 135

[CT01] D. Caseiro and I. Trancoso, "Transducer composition for "on-the-fly" lexicon and language model integration," in *Proc. ASRU*, 2001, pp. 393–396. DOI: 10.1109/ASRU.2001.1034667 95, 99

[CT03] D. Caseiro and I. Trancoso, "A tail-sharing WFST composition for large vocabulary speech recognition," in *Proc. ICASSP*, vol. I, 2003, pp. 356–359. DOI: 10.1109/ICASSP.2003.1198791 95

[CT06] D. Caseiro and I. Trancoso, "A specialized on-the-fly algorithm for lexicon and language model composition," *IEEE Transactions on Audio, Speech, and Language Processing*, vol. 14, no. 4, pp. 1281–1291, 2006. DOI: 10.1109/TSA.2005.860838 99

[CTOV02] D. Caseiro, L. Trancoso, L. Oliveira, and C. Viana, "Grapheme-to-phone using finite-state transducers," in *Proc. IEEE Workshop on Speech Synthesis*, 2002, pp. 215–218. DOI: 10.1109/WSS.2002.1224412 135

[DDCTOSF07] P.R. Dixon, D. Caseiro, T. Oonishi, and S. Furui, "The TITECH large vocabulary WFST speech recognition system," in *Proc. ASRU*, 2007, pp. 443–448. DOI: 10.1109/ASRU.2007.4430153 136

[DH01] H. J. G. A. Dolfing and I. L. Hetherington, "Incremental language models for speech recognition using finite-state transducers," in *Proc. ASRU*, 2001, pp. 194–197. DOI: 10.1109/ASRU.2001.1034620 95, 110

[DHK12] P. R. Dixon, C. Hori, and H. Kashioka, "A comparison of dynamic WFST decoding approaches," in *Proc. ICASSP*, Kyoto, Japan, 2012, pp. 4209–4212. DOI: 10.1109/ICASSP.2012.6288847 93, 126

[Eis02] J. Eisner, "Parameter estimation for probabilistic finite-state transducers," in *Proc. ACL*, 2002, pp. 1–8. DOI: 10.3115/1073083.1073085 133

[FBP88] W. M. Fisher, J. Bernstein, and D. S. Pallett, "The DARPA 1000-word resource management database for continuous speech recognition," in *Proc. ICASSP*, vol. 1, 1988, pp. 651–654. DOI: 10.1109/ICASSP.1988.196669 1

[FMI00] S. Furui, K. Maekawa, and H. Isahara, "A Japanese national project on spontaneous speech corpus and processing technology," in *Proc. of ASR*, 2000, pp. 244–248. 83

[FSM]   "AT&T FSM Library," web page http://www.itl.nist.gov/iad/mig/
          tests/rt/2009/index.html. 136

[Fur86]   S. Furui, "Speaker-independent isolated word recognition using dynamic features
          of speech spectrum," *IEEE Transactions on Acoustics, Speech, and Signal Processing*,
          vol. 34, no. 1, pp. 52–59, 1986. DOI: 10.1109/TASSP.1986.1164788 12

[Gar08]   P. Garner, "Silence models in weighted finite-state transducers," in *Proc. Interspeech*,
          Brisbane, Australia, 2008, pp. 1817–1820. 75

[GBM95]   P. S. Gopalakrishnan, L. R. Bahl, and R. L. Mercer, "A tree search strategy for
          large-vocabulary continuous speech recognition," in *Proc. ICASSP*, vol. 572–575,
          1995. DOI: 10.1109/ICASSP.1995.479662 4

[Goo53]   I. J. Good, "The population frequencies of species and the estimation of population
          parameters," *Biometrika*, vol. 40, no. 3-4, pp. 237–264, 1953.
          DOI: 10.2307/2333344 22

[HAH01]   X. Huang, A. Acero, and H.-W. Hon, *Spoken Language Processing: A Guide to
          Theory, Algorithm, and System Development*.   Prentice Hall, 2001. 12

[HES00]   H. Hermansky, D. P. W. Ellis, and S. Sharma, "Tandem connectionist feature
          extraction for conventional HMM systems," in *Proc. ICASSP*, vol. 3, 2000, pp.
          1635–1638. DOI: 10.1109/ICASSP.2000.862024 12

[Het04]   I. L. Hetherington, "The MIT finite-state transducer toolkit for speech and lan-
          guage processing," in *Proc. Interspeech—ICSLP*, 2004. 136

[Het05]   I. L. Hetherington, "A multi-pass, dynamic-vocabulary approach to real-time,
          large-vocabulary speech recognition," in *Proc. Interspeech—Eurospeech*, 2005, pp.
          545–548. 131

[HGD90]   C. T. Hemphill, J. J. Godfrey, and G. R. Doddington, "The ATIS spoken language
          systems pilot corpus," in *DARPA Speech and Natural Language Workshop*, Hidden
          Valley, Pennsylvania, June 1990. DOI: 10.3115/116580.116613 1

[HHM03]   T. Hori, C. Hori, and Y. Minami, "Speech summarization using weighted finite-
          state transducers," in *Proc. Eurospeech*, 2003, pp. 2817–2820. 134

[HHM04]   T. Hori, C. Hori, and Y. Minami, "Fast on-the-fly composition for weighted finite-
          state transducers in 1.8 million-word vocabulary continuous speech recognition,"
          in *Proc. Interspeech—ICSLP*, vol. 1, 2004, pp. 289–292. 6, 95, 110

[HHMN07] T. Hori, C. Hori, Y. Minami, and A. Nakamura, "Efficient WFST-based one-pass decoding with on-the-fly hypothesis rescoring in extremely large vocabulary continuous speech recognition," *IEEE Transactions on Audio, Speech, and Language Processing*, vol. 15, no. 4, pp. 1352–1365, 2007. DOI: 10.1109/TASL.2006.889790 6, 95, 110

[HHR+12] B. Hoffmeister, G. Heigold, D. Rybach, R. Schluter, and H. Ney, "WFST enabled solutions to ASR problems: Beyond HMM decoding," *IEEE Transactions on Audio, Speech, and Language Processing*, vol. 20(2), pp. 551–564, 2012. DOI: 10.1109/TASL.2011.2162402 132

[HMU06] J. E. Hopcroft, R. Motwani, and J. D. Ullman, *Introduction to Automata Theory, Languages, and Computation*, 3rd ed. Addison-Wesley Publishing Company, 2006. 4, 45, 75

[HN05] T. Hori and A. Nakamura, "Generalized fast on-the-fly composition algorithm for WFST-based speech recognition," in *Proc. Interspeech—Eurospeech*, 2005, pp. 557–560. 94, 95, 110, 130

[HOM+08] C. Hori, K. Ohtaki, T. Misu, H. Kashioka, and S. Nakamura, "Dialog management using weighted finite-state transducers," in *Proc. Interspeech*, 2008, pp. 211–214. DOI: 10.1109/ASRU.2009.5373350 136

[HOM+09] C. Hori, K. Ohtaki, T. Misu, H. Kashioka, and S. Nakamura, "Statistical dialog management applied to WFST-based dialog systems," in *Proc. ICASSP*, 2009, pp. 4793–4796. DOI: 10.1109/ICASSP.2009.4960703 136

[HSG06] I. L. Hetherington, H. Shu, and J. R. Glass, "Flexible multi-stream framework for speech recognition using multi-tape finite-state transducers," in *Proc. ICASSP*, 2006, pp. 417–420. DOI: 10.1109/ICASSP.2006.1660046 132

[HSN09] G. Heigold, R. Schluter, and H. Ney, "A multimedia retrieval system using speech input," in *Proc. ICASSP*, 2009, pp. 3749–3752. DOI: 10.1145/1647314.1647356 132

[HWM03a] T. Hori, D. Willett, and Y. Minami, "Language model adaptation using WFST-based speaking-style translation," in *Proc. ICASSP*, vol. I, 2003, pp. 228–231. DOI: 10.1109/ICASSP.2003.1198759 131

[HWM03b] T. Hori, D. Willett, and Y. Minami, "Paraphrasing spontaneous speech using weighted finite-state transducers," in *Proc. SSPR*, 2003. 134

[HZN09] G. Heigold, G. Zweig, and P. Nguyen, "A flat direct model for speech recognition," in *Proc. ICASSP*, 2009, pp. 3861–3864. DOI: 10.1109/ICASSP.2009.4960470 133

[IOR94]   R. Iyer, M. Ostendorf, and J. R. Rohlicek, "Language modeling with sentence-level mixtures," in *Proc. Workshop on Human Language Technology*, 1994, pp. 82–87. DOI: 10.3115/1075812.1075828 131

[JBM75]   F. Jelinek, L. R. Bahl, and R. L. Mercer, "Design of a linguistic statistical decoder for the recognition of continuous speech," *IEEE Transactions on Information Theory*, vol. IT-21, no. 3, pp. 250–256, 1975. DOI: 10.1109/TIT.1975.1055384 9

[Jel98]   F. Jelinek, Ed., *Statistical Methods for Speech Recognition*. The MIT Press, 1998. 1, 9

[JMSS91] F. Jelinek, B. Merialdo, R. S., and M. Strauss, "A dynamic language model for speech recognition," in *Proc. DARPA Workshop on Speech and Natural Language*, 1991, pp. 293–295. DOI: 10.3115/112405.112464 131

[JOIF09] A. T. Jensson, T. Oonishi, K. Iwano, and S. Furui, "Development of a WFST based speech recognition system for a resource deficient language using machine translation," in *Proc. APSIPA ASC*, 2009, pp. 50–56. 131

[JSHS03] J. J. Schalkwyk, I. L. Hetherington, and E. Story, "Speech recognition with dynamic grammars using finite-state transducers," in *Proc. Eurospeech*, 2003, pp. 1969–1972. 131

[KA98]   N. Kumar and H. G. Andreou, "Heteroscendastic discriminant analysis and reduced rank HMMs for improved speech recognition," *Speech Communication*, vol. 26, pp. 283–297, 1998. DOI: 10.1016/S0167-6393(98)00061-2 12

[Kat87]   S. M. Katz, "Estimation of probabilities from sparse data for the language model component of a speech recognizer," *IEEE Transactions on Acoustics, Speech, and Signal Processing*, vol. 35, no. 3, pp. 400–401, 1987. DOI: 10.1109/TASSP.1987.1165125 22

[KHG⁺91] P. Kenny, R. Hollan, V. Gupta, M. Lennig, P. Mermelstein, and D. O'Shaughnessy, "A*-admissible heuristics for rapid lexical access," in *Proc. ICASSP*, 1991, pp. 689–692. DOI: 10.1109/ICASSP.1991.150433 4

[KLNB95] R. Kuhn, A. Lazarides, Y. Normandin, and J. Brousseau, "Improved decision trees for phonetic modeling," in *Proc. ICASSP*, vol. 1, 1995, pp. 552–555. DOI: 10.1007/11965152_26 131

[KN04]   S. Kanthak and H. Ney, "FSA: An efficient and flexible C++ toolkit for finite state automata using on-demand computation," in *Proc. ACL*, 2004, pp. 510–517. 136

[KWHN12] Y. Kubo, S. Watanabe, T. Hori, and A. Nakamura, "Structural classification methods based on weighted finite-state transducers for automatic speech recognition," *IEEE Transactions on Audio, Speech, and Language Processing*, 2012, in press. 133

[Lee88] K.-F. Lee, "Large-vocabulary speaker-independent continuous speech recognition: the SPHINX system," *PhD thesis, Carnegie Mellon University*, April 1988. 17

[LGHW10] X. Liu, M. J. F. Gales, J. L. Hieronymus, and P. C. Woodland, "Language model combination and adaptation using weighted finite state transducers," in *Proc. ICASSP*, 2010, pp. 5390–5393. 131

[LJ93] L. Labiner and B.-H. Juang, *Fundamentals of Speech Recognition*. Prentice Hall, 1993. 12

[Low76] B. Lowerre, "The HARPY speech recognition system," *PhD theses, Dept. of Computer Science, Carnegie-Mellon University, Pittsburgh, PA, USA*, 1976. 3, 27

[LPR99] A. Ljolje, F. Pereira, and M. Riley, "Efficient general lattice generation and rescoring," in *Proc. Eurospeech*, 1999, pp. 1251–1254. 88, 94, 126

[LS10] M. Lehr and I. Shafran, "Discriminatively estimated joint acoustic, duration, and language model for speech recognition," in *Proc. ICASSP*, 2010, pp. 5542–5545. DOI: 10.1109/ICASSP.2010.5495227 133

[MDH09] A. Mohamed, G. Dahl, and G. Hinton, "Deep belief networks for phone recognition," in *Proc. NIPS Workshop on Deep Learning for Speech Recognition*, 2009. 16

[MDMD⁺06] D. Moore, J. Dines, M. Magimai Doss, J. Vepa, O. Cheng, and T. Hain, "Juicer: a weighted finite-state transducer decoder," *Machine Learning for Multimodal Interaction, Lecture Notes in Computer Science*, vol. 4299, pp. 285–296, 2006. 136

[MHLR⁺07] E. McDermott, T. J. Hazen, J. Le Roux, A. Nakamura, and S. Katagiri, "Discriminative training for large vocabulary speech recognition using minimum classification error," *IEEE Transactions on Audio, Speech and Language Processing*, vol. 15, pp. 203–223, 2007. DOI: 10.1109/TASL.2006.876778 132

[MIT] "The MIT FST Toolkit," web page http://people.csail.mit.edu/ilh/fst. 136

[Moh02] M. Mohri, "Generic epsilon-removal and input epsilon-normalization algorithms for weighted transducers," *International Journal of Foundations of Computer Science*, vol. 13(1), pp. 129–143, 2002. DOI: 10.1142/S0129054102000996 65

[Moh09] M. Mohri, "Weighted automata algorithms," in *Handbook of Weighted Automata*, M. Droste, W. Kuich, and H. Vogler, Eds. Springer-Verlag New York Inc., 2009. DOI: 10.1007/978-3-642-01492-5 56, 57, 61, 65

[MPR96] M. Mohri, F. Pereira, and M. Riley, "Weighted automata in text and speech processing," in *Proc. ECAI-96, Workshop on Extended Finite State Models of Language*, Budapest, Hungary, 1996. 4

[MPR00] M. Mohri, F. Pereira, and M. Riley, "The design principles of a weighted finite-state transducer library," *Theoretical Computer Science*, vol. 231(1), pp. 17–32, 2000. DOI: 10.1016/S0304-3975(99)00014-6 136

[MPR02] M. Mohri, F. Pereira, and M. Riley, "Weighted finite-state transducers in speech recognition," *Computer Speech and Language*, vol. 16, pp. 69–88, 2002. DOI: 10.1006/csla.2001.0184 4, 41, 71, 80, 83, 94, 95

[MR97] M. Mohri and M. Riley, "Weighted determinization and minimization for large vocabulary speech recognition," in *Proc. Eurospeech*, vol. 1, 1997, pp. 131–134. 95

[MR01] M. Mohri and M. Riley, "A weight pushing algorithm for large vocabulary speech recognition," in *Proc. Eurospeech*, 2001, pp. 1603–1606. 91

[MSK07] J. McDonough, E. Stoimenov, and D. Klakow, "An algorithm for fast composition of weighted finite-state transducers," in *Proc. ASRU*, 2007, pp. 461–466. DOI: 10.1109/ASRU.2007.4430156 95, 99

[Ney84] H. Ney, "The use of a one-stage dynamic programming algorithm for connected word recognition," *IEEE Transactions on Acoustics, Speech, and Signal Processing*, vol. ASSP-32, no. 2, pp. 263–271, Apr. 1984. DOI: 10.1109/TASSP.1984.1164320 3

[NHUTO92] H. Ney, R. Haeb-Umbach, B. Tran, and M. Oerder, "Improvements in beam search for 10000-word continuous speech recognition," in *Proc. ICASSP*, vol. I, 1992, pp. 9–12. DOI: 10.1109/89.279287 27

[ODIF09a] T. Oonishi, P. R. Dixon, K. Iwano, and S. Furui, "Generalization of specialized on-the-fly composition," in *Proc. ICASSP*, 2009, pp. 4317–4320. DOI: 10.1109/ICASSP.2009.4960584 95, 99, 102

[ODIF09b] T. Oonishi, P. R. Dixon, K. Iwano, and S. Furui, "Optimization of on-the-fly composition for WFST-based speech recognition decoders," *IEICE Transactions on Information and Systems*, vol. J92-D, no. 7, pp. 1026–1035, 2009, (in Japanese). 102, 103

[OHIN11] T. Oba, T. Hori, A. Ito, and A. Nakamura, "Round-robin duel discriminative language models in one-pass decoding with on-the-fly error correction," in *Proc. ICASSP*, 2011, pp. 5588–5591. DOI: 10.1109/ICASSP.2011.5947626 131

[ONA97] S. Ortmanns, H. Ney, and X. Aubert, "A word graph algorithm for large vocabulary continuous speech recognition," *Computer Speech and Language*, vol. 1, pp. 43–72, 1997. DOI: 10.1006/csla.1996.0022 4, 33, 38, 94

[ONE96] S. Ortmanns, H. Ney, and A. Eiden, "Language-model look-ahead for large vocabulary speech recognition," in *Proc. ICSLP*, 1996, pp. 2095–2098. DOI: 10.1109/ICSLP.1996.607215 103

[OPE] "OpenFst Library," web page http://www.openfst.org/twiki/bin/view/FST/WebHome. 136

[Pal89] D. S. Pallett, "Benchmark tests for DARPA resource management database performance evaluations," in *Proc. ICASSP*, 1989, pp. 536–539. DOI: 10.1109/ICASSP.1989.266482 1

[Pau91] D. B. Paul, "Algorithm for an optimal A* search and linearizing the search in the stack decoder," in *Proc. ICASSP*, 1991, pp. 693–696. DOI: 10.1109/ICASSP.1991.150434 4

[PBA+11] D. Povey, L. Burget, M. Agarwal, P. Akyazi, F. Kai, A. Ghoshal, O. Glembek, N. Goel, M. Karafiat, A. Rastrow, R. C. Rosei, P. Schwarz, and S. Thomas, "The subspace Gaussian mixture model—A structured model for speech recognition," *Computer Speech & Language*, vol. 25, no. 2, pp. 404–439, April 2011. DOI: 10.1016/j.csl.2010.06.003 16

[PFFG90] D. S. Pallett, W. M. Fisher, J. G. Fiscus, and J. S. Garofolo, "DARPA ATIS test results June 1990," in *Proc. Speech and Natural Language Workshop*, R. Stern, Ed. Morgan Kaufmann Publishers, Inc., June 1990, pp. 114–121. 1

[PGB+11] D. Povey, A. Ghoshal, G. Boulianne, L. Burget, O. Glembek, N. Goel, M. Hannemann, P. Motlicek, Y. Qian, P. Schwarz, J. Silovsky, G. Stemmer, and K. Vesely, "The Kaldi speech recognition toolkit," in *Proc. ASRU*, 2011. 136

[PR96] F. Pereira and M. Riley, "Speech recognition by composition of weighted finite automata," in *Finite-State Language Processing*. MIT Press, 1996, pp. 431–453. 4

[PRS94] F. Pereira, M. Riley, and R. Sproat, "Weighted rational transductions and their application to human language processing," in *Proc. ARPA Workshop on Human Language technology*, 1994, pp. 249–254. DOI: 10.3115/1075812.1075870 4

[PW02] D. Povey and P. C. Woodland, "Minimum phone error and I-smoothing for improved discriminative training," in *Proc. ICASSP*, vol. I, 2002, pp. 105–108. DOI: 10.1109/ICASSP.2002.5743665 13, 132

[Rev92] D. Revuz, "Minimisation of acyclic deterministic automata in linear time," *Theoretical Computer Science*, vol. 92(1), pp. 181–189, 1992. DOI: 10.1016/0304-3975(92)90142-3 64

[RHL+11] D. Rybach, S. Hahn, P. Lehnen, D. Nolden, M. Sundermeyer, Z. Tuske, S. Wiesler, R. Schluter, and N. Ney, "RASR—The RWTH Aachen university open source speech recognition toolkit," in *Proc. ASRU*, 2011. 136

[RMB+94] S. Renals, N. Morgan, H. Boulard, M. Cohen, and H. Franco, "Connectionist probability estimators in HMM speech recognition," *IEEE Transactions on Speech and Audio Processing*, vol. 2, no. 1, pp. 161–174, 1994. DOI: 10.1109/89.260359 16

[RPM97] M. Riley, F. Pereira, and M. Mohri, "Transducer composition for context-dependent network expansion," in *Proc. Eurospeech*, 1997, pp. 1427–1430. 71

[RS97] E. Roche and Y. Schabes, Eds., *Finite-State Language Processing*.   A Bradford Book, 1997. 44

[RSCJ04] B. Roark, M. Saraclar, M. Collins, and M. Johnson, "Discriminative language modeling with conditional random fields and the perceptron algorithm," in *Proc. ACL*, 2004. DOI: 10.3115/1218955.1218962 130

[RSN12] D. Rybach, R. Schluter, and H. Ney, "Silence is golden: modeling non-speech events in WFST-based dynamic network decoders," in *Proc. ICASSP*, Kyoto, Japan, 2012, pp. 4205–4208. DOI: 10.1109/ICASSP.2012.6288846 75

[RWT] "The RWTH FSA Toolkit," web page http://www-i6.informatik.rwth-aachen.de/~kanthak/fsa.html. 136

[SA90] R. Schwartz and Y. Austin, "A comparison of several approximate algorithms for finding multiple (N-best) sentence hypotheses," in *Proc. ICASSP*, 1990, pp. 701–704. DOI: 10.1109/ICASSP.1991.150436 39

[SC70] H. Sakoe and S. Chiba, "A similarity evaluation of speech patterns by dynamic programming (in japanese)," in *the Dig. 1970 Nat. Meeting*, Inst. Electrn. Comm. Eng. Japan, July 1970, p. 136. 2

[SC71] H. Sakoe and S. Chiba, "A dynamic programming approach to continuous speech recognition," in *Proc. ICA*, Budapest, Hungary, Paper 20 C 13, August 1971, pp. 65–68. 2

[SC90]   R. Schwartz and Y. Chow, "The N-best algorithm: an efficient and exact procedure for finding the N most likely sentence hypotheses," in *Proc. ICASSP*, 1990, pp. 81–84. DOI: 10.1109/ICASSP.1990.115542 4

[Sch00]   M. Schuster, "Memory-efficient LVCSR search using a one-pass stack decoder," *Computer Speech & Language*, vol. 14(1), pp. 47–77, January 2000. DOI: 10.1006/csla.1999.0135 4

[SH91]   F. K. Soong and E.-F. Huang, "A tree-trellis based fast search for finding the N-best sentence hypotheses in continuous speech recognition," in *Proc. ICASSP*, vol. 1, 1991, pp. 705–708. DOI: 10.1109/ICASSP.1991.150437 4

[SH02]   H. Shu and I. L. Hetherington, "EM training of finite-state transducers and its application to pronunciation modeling," in *Proc. ICSLP*, 2002, pp. 1293–1296. 132

[SH05]   M. Schuster and T. Hori, "Efficient generation of high-order context-dependent weighted finite state transducers for speech recognition," in *Proc. ICASSP*, 2005, pp. 201–204. DOI: 10.1109/ICASSP.2005.1415085 132

[SLY11]   F. Seide, G. Li, and D. Yu, "Conversational speech transcription using context-dependent deep neural networks," in *Proc. Interspeech*, 2011, pp. 437–440. 16

[SM06]   E. Stoimenov and J. McDonough, "Modeling polyphone context with weighted finite-state transducers," in *Proc. ICASSP*, vol. I, 2006, pp. 121–124. DOI: 10.1109/ICASSP.2006.1659972 132

[Spr96]   R. Sproat, "Multilingual text analysis for text-to-speech synthesis," in *Proc. ICSLP*, vol. 3, 1996, pp. 1365–1368. DOI: 10.1017/S1351324997001654 135

[Sto98]   A. Stolcke, "Entropy-based pruning of backoff language models," in *Proc. DARPA Broadcast News Transcription and Understanding Workshop*, 1998, pp. 270–274. 78

[SW00]   K. Shinoda and T. Watanabe, "MDL-based context-dependent subword modeling for speech recognition," *Acoustic Science and Technology*, vol. 21, no. 2, pp. 79–86, 2000. DOI: 10.1250/ast.21.79 19

[SYM⁺96]   T. Shimizu, H. Yamamoto, H. Masataki, S. Matsunaga, and Y. Sagisaka, "Spontaneous dialogue speech recognition using cross-word context constrained word graphs," in *Proc. ICASSP*, 1996, pp. 145–148. DOI: 10.1109/ICASSP.1996.540311 126

[WHMN10]   S. Watanabe, T. Hori, E. McDermott, and A. Nakamura, "A discriminative model for continuous speech recognition based on weighted finite state transducers," in *Proc. ICASSP*, 2010, pp. 4922–4925. DOI: 10.1109/ICASSP.2010.5495096 133

[WMMK01] D. Willett, E. McDermott, Y. Minami, and S. Katagiri, "Time and memory efficient Viterbi decoding for LVCSR using a precompiled search network," in *Proc.
Eurospeech*, 2001, pp. 847–850. 95, 110, 111

[WMNU04] S. Watanabe, Y. Minami, A. Nakamura, and N. Ueda, "Variational Bayesian estimation and clustering for speech recognition," *IEEE Transactions on Speech and
Audio Processing*, vol. 12, pp. 365–381, 2004. DOI: 10.1109/TSA.2004.828640 19

[YOW94] S. J. Young, J. J. Odell, and P. C. Woodland, "Tree-based state tying for high accuracy acoustics modeling," in *Proc. ARPA Human Language Technology Workshop*,
1994, pp. 307–312. DOI: 10.3115/1075812.1075885 18

[ZLCJ12] Y. Zhao, A. Ljolje, D. Caseiro, and B.-H. Juang, "A general discriminative training
algorithm for speech recognition using weighted finite-state transducers," in *Proc.
ICASSP*, Kyoto, Japan, 2012, pp. 4217–4220.
DOI: 10.1109/ICASSP.2012.6288849 132

[ZN09] G. Zweig and P. Nguyen, "A segmental CRF approach to large vocabulary continuous speech recognition," in *Proc. ASRU*, 2009, pp. 152–157.
DOI: 10.1109/ASRU.2009.5372916 133

Authors' Biographies

TAKAAKI HORI

Takaaki Hori received the B.E. and M.E. degrees in electrical and information engineering from Yamagata University, Yonezawa, Japan, in 1994 and 1996, respectively, and a Ph.D. degree in system and information engineering from Yamagata University in 1999.

Since 1999, he has been engaged in research on spoken language processing at the Cyber Space Laboratories, Nippon Telegraph, and Telephone (NTT) Corporation, Kyoto, Japan. He was a visiting scientist at the Massachusetts Institute of Technology, Cambridge, from 2006 to 2007. He is currently a senior research scientist in the NTT Communication Science Laboratories, NTT Corporation.

He received the 22nd Awaya Prize Young Researcher Award from the Acoustical Society of Japan (ASJ) in 2005, the 24th TELECOM System Technology Award from the Telecommunications Advancement Foundation in 2009, and the IPSJ Kiyasu Special Industrial Achievement Award from the Information Processing Society of Japan in 2012.

He is a member of Institute of Electrical and Electronic Engineers (IEEE), the Institute of Electronics, Information, and Communication Engineers (IEICE), and the ASJ.

ATSUSHI NAKAMURA

Atsushi Nakamura received the B.E., M.E., and Dr.Eng. degrees from Kyushu University, Fukuoka, Japan, in 1985, 1987, and 2001, respectively.

In 1987, he joined Nippon Telegraph and Telephone Corporation (NTT), where he engaged in the research and development of network service platforms, including studies on application of speech processing technologies into network services, at Musashino Electrical Communication Laboratories, Tokyo, Japan. From 1994 to 2000, he was with Advanced Telecommunications Research (ATR) Institute, Kyoto, Japan, as a Senior Researcher, working on the research of spontaneous speech recognition, construction of spoken language database, and development of speech translation systems. Since April 2000, he has been with NTT Communication Science Laboratories, Kyoto, Japan, and is currently the head of Signal Processing Research Group.

Dr. Nakamura is a senior member of the Institute of Electrical and Electronic Engineers (IEEE), and serves or served as a member of the IEEE Machine Learning for Signal Processing (MLSP) Technical Committee, a Vice Chair of the IEEE Signal Processing Society Kansai Chapter, etc. He is also a member of the Institute of Electronics, Information and Communication Engineering (IEICE) and the Acoustical Society of Japan (ASJ). He received the IEICE Paper Award in 2004, and received twice the TELECOM System Technology Award of the Telecommunications Advancement Foundation, in 2006 and 2009.

Printed in the United States
by Baker & Taylor Publisher Services